CONCRETE AND MASONRY

Techniques and Design

CONCRETE AND MASONRY

Techniques and Design

R. J. De Cristoforo

RESTON PUBLISHING COMPANY, INC.
Reston, Virginia 22090
A Prentice-Hall Company

Library of Congress Cataloging in Publication Data

De Cristoforo, R J
    Concrete and masonry.

    1. Concrete construction.   2. Masonry. I. Title.
TA682.42.D42   624'.1834   74-31193
ISBN 0-87909-149-5

10 9 8 7 6 5 4 3 2 1

Printed in the United States of America.

Contents

Preface

Concrete and masonry work can be challenging and rewarding regardless of whether the involvement is vocational or avocational. Today's professional artisans are enjoying monetary returns that are justified by their necessary skills. These skills are more in demand than ever before because of the inroads made by the subject materials into all phases of structural and decorative design. The good mason is as necessary as a lawyer or a banker or an electronics engineer.

A basic difference between the professional mason and the amateur is the push-button knowledge that should be part of the pro's tool kit. Because of constant practice and learning, the vocationalist acquires proficiency that eliminates the prelude the amateur must face.

Often though, it may be difficult to judge the status of the worker by viewing the end result. In essence, it really doesn't make much difference whether the successful job was accomplished by one who moved quickly because of past experience or by one who had an instruction sheet (or an instructor) at his side through each phase of the project.

In either case, dedication and pride of craftsmanship are as important as correct procedures. The work of the good amateur is as distinguishable as the work of the good professional.

It is my belief that the vision of the complete job as seen by, for example, a landscape architect is as important to the craftsman/contributor as it is to the designer. It's much too easy to feel lost in a confined area. The reason for a poor or indifferent job can often be traced to a tunnel vision that results when the importance of the detail is overshadowed by the whole.

The thought applies whether the worker is a student, an employee, a contractor, or a layman/homeowner who plays all parts. It is true of the individual project and of designs that result from multiple, contributing elements. Each of the elements, of course, is a project in itself and is composed of its own details.

This book makes an attempt to broaden the viewpoint of the worker so he can take pride in both his contribution to a scheme and in his craftsmanship. It goes beyond the techniques of doing by suggesting and detailing applications. There isn't much fun in learning to drive if you can't go anyplace.

I am grateful to the following organizations for their cooperation and permission to include some of their illustrations in this book.

PORTLAND CEMENT ASSOCIATION

BRICK INSTITUTE OF AMERICA

GOLDBLATT TOOL CO.

FISCHER OF AMERICA, INC.

SAKRETE, INC.

STANDARD DRY WALL PRODUCTS, INC.

CON-FORM EQUIPMENT CORPORATION

THE STANLEY WORKS

FLEMING MANUFACTURING CO.

PLASTRONICS, INC.

ELDORADO STONE CORPORATION

SUPERIOR CONCRETE ACCESSORIES, INC.

BOMANITE CORPORATION

Form, Function, and Practice

It is impossible to view the existing environment without being aware of the myriad applications of concrete and masonry products. Because of these materials, we can ride and walk safely, be housed more comfortably, swim in a backyard, view a private waterfall, attain security and privacy and, generally, achieve a higher degree of well-being.

Few materials are so flexible inside and outside the home and in such harmony with nature's own products. The techniques of design and application have progressed to the point where they can be made artistically compatible with either modern or traditional interpretations. Walls, walks, and houses, terraces and planters, posts and benches—all can be designed and textured for maximum function and durability and minimum intrusion on a natural scene.

It is true that many of the designers of yesterday's commercial projects were less than sensitive to the environmental and visual impact of acres of bare concrete. The same was true of projects around the home. A terrace can be constructed of uninviting square yards of concrete, or it can be pleasant and restful, without sacrificing practicality, through a design that includes wooden grids, open squares for shrubs, texture, border greenery, and trees at the perimeter.

The difference between poor design and good design lies principally in a plan that visualizes the whole: The house and lot are viewed as an ensemble

FIG. 1–1   This graceful concrete curb that separates a sloping lawn from a graveled area demonstrates the flexibility of the material. The forming requires time but the results are worth it.

FIG. 1–2   An overall plan, made before the first shovelful of mix is prepared, will help produce good end results. The work is still done project by project but all will be compatible in the master scheme.

that involves leisure, traffic, and utility areas. Obviously, each element requires individual consideration.

To get off to a good start, nothing is more essential than a scaled drawing of the lot and the house, regardless of whether the structure is new and not yet landscaped or is established and thus partially or wholly landscaped. The drawing doesn't have to be more than a bird's-eye view of perimeter lines, with all entries indicated—that is, the main garage door and any smaller door that permits access from the garage to an adjacent area, the front entry door, and all side and rear doors.

Entryways do much to establish traffic patterns, and walks and paths should be designed to coordinate with them. For example, a utility door in a garage that contains laundry facilities or that is adjacent to a separate inside utility room should exit onto a practical walk that leads to an outside clothesline, the trash-can area, garden tool shed, and so on. The width of the walk should just accommodate a single person who might be carrying a clothes basket or pushing a wheelbarrow. On the other hand, an entry walk from a driveway or sidewalk to the front door deserves more consideration as to design and should be wide enough to accommodate two people walking abreast.

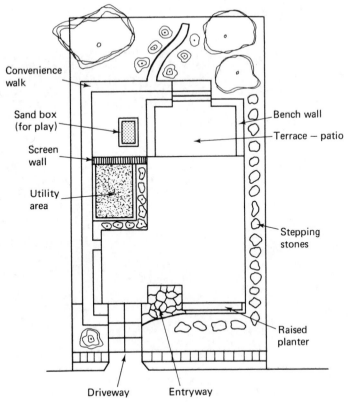

FIG. 1-3   The many faces of concrete and masonry work.

A single-car garage won't accommodate a two-car driveway, but when space permits, parking areas should be allowed for in the design. Even when space is at a premium and driveway width must be limited, landing strips can be included so that people exiting from automobiles will have sure, clean footing.

It can be seen that practical considerations have much to do with design. Something attractive but inconvenient and difficult to maintain is not good design—neither is a wide driveway slab that is too thin to hold up under automobile traffic. This is why a scaled drawing is so important: It will permit an objective analysis so that the whole may be viewed in terms of livability as well as appearance.

Naturally, as this book will demonstrate, function is not the sole design factor. A walk from A to B may be a practical necessity, but it doesn't have to be prosaic. For example, concrete and masonry materials are quite varied; a wise choice in this area can make the difference between a merely adequate job and one that provides pride of accomplishment. The latter is a commendable goal whether the worker is a novice or a professional. Since any job requires time, effort, and money, it seems logical to work toward a better-than-acceptable result if the effort is to be made at all.

Being an amateur is often an advantage. Although your time should not be considered expendable, it is nevertheless a contribution with which you can

FIG. 1–4   A low concrete wall can be used as a seat, especially when a wooden bench top is installed. Note the exposed aggregate on the wall—done by hosing and brushing as soon as forms are removed.

FIG. 1-5   Wise selection and imaginative use of standard materials contribute to outstanding design. Here, "rocket block" and "split block" combine beautifully with wooden members and furniture. (Besser Co. Photo)

afford to be more generous. Frequently, it is extra time that results in the difference between "good" and "super." For example, a formal brick-floored patio will be more successful if joint lines are uniform. This can be attempted by eye or by using temporary, small pieces of plywood as spacers. It is logical to assume that the plywood-gauge method will result in more precise work, no matter who is doing it.

Of course, time is not the sole ingredient in successful concrete and masonry work. Good materials, correct proportions, proper mixing, careful placement, and the like—all contribute to a good result. But these ingredients are based for the most part on techniques and knowledge available to anyone.

On the other hand, time should not be wasted. Time allotted should be proportional to the job or the job phase in question. For example, temporary concrete forms are critical but they do not have to be pretty. They must be structurally capable of containing the great mass and weight of a concrete pour, but they do not require the touch of a master cabinetmaker. To make an analogy, much less time need be devoted to a nailed butt joint than to a glued miter.

Geographical location is a crucial factor in concrete and masonry work. Local building codes set needed standards of safety and legal responsibility; it is thus imperative that all structural projects conform to these codes. Too often, the amateur, especially, is guilty of nonconformity in this area; this is a two-way error. For one thing, projects carried out without regard to local standards can be unsafe and illegal; second, the worker who chooses to ignore the rules is passing up a tremendous source of technical advice that relates especially to his area. In essence, by conforming to these codes, you will be well on the road to professional results and will avoid all the mistakes that led to the establishment of building codes in the first place. The importance of codes cannot be over-emphasized; we shall caution you about them frequently.

One consideration that the amateur must face is the amount of concrete he can handle in a given period of time—particularly with regard to pours for such projects as foundations, large retaining walls, extensive slabs, and the like. Factors that can mean the difference between success and disaster are these:

> Be sure that the site is ready for the pour. This involves a recheck of the subbase and all formwork.

> Have a clear picture of what must be done once the pouring starts. A "dry" run (to practice) won't hurt.

> Have an assistant (a neighbor or family member) who has been briefed on the procedure.

The amateur who wishes to avoid the chore of handling a large amount of concrete can do all the work that precedes the pour itself and then call in pro-

FIG. 1–6   Being prepared and knowing exactly what must be done are the secrets of a non-frustrating concrete pour. Here, one worker uses a jitterbug tamper while the other follows with a long-handled float.

fessional help. In other words, he undertakes the designing, the subbase preparation, and the forming, and enjoys the creativity and the substantial reduction in cost that result.

In the final analysis there is rarely a good excuse for poor work, but there can be reasons. The most common is an impulse to follow procedures haphazardly in an effort to cut corners and so save time. On the other hand, good concrete and masonry work that will last indefinitely can be accomplished by anyone who is dedicated enough to do the job in a technically correct fashion.

FIG. 1–7   Flexibility in choice of materials applies to products available for outdoor floors. These pre-shaped "Patio Pavers" may be placed on a suitable sand bed or over a subbase of concrete.

QUESTIONS

1–1   How have concrete and masonry materials contributed to our well-being?
1–2   Name some of the intriguing aspects of concrete and masonry materials?
1–3   Give examples of what can be done to minimize the impact of large slabs of concrete.

FIG. 1–8   Quite often the real thing can be combined with an imitator. Here, the wall and hearth top are real brick, while the hearth facings are covered with plastic—a good procedure when weight must be kept to a minimum.

1–4   Why is it wise to have an overall plan of the house and lot?
1–5   What are traffic patterns (around a home)?
1–6   Why are the traffic patterns important in relation to design?
1–7   Why is it important and why does it make good sense to work in accord with local building codes?
1–8   List three considerations, in relation to a concrete pour, that can mean the difference between success and failure.

2

The Fluid Stone

All concrete has a certain sameness of appearance that makes it difficult to distinguish between a good and a mediocre job until sufficient time has elapsed to test the material's quality. By then, of course, it is too late. Short of living with what will probably be a nuisance, the only remedial action is to literally break up the material for removal and to replace it with a good product.

Rigid standards control the quality of the materials that go into concrete. The causes of failure thus stem mostly from poor workmanship and penny-saving experiments. Ill-advised shortcuts and lackadaisical craftsmanship don't make sense when the product, honestly executed, will last and be attractive for many lifetimes.

A prime reason for bad concrete is incorrect proportioning of the materials required for a mix. The amounts required are as specific, and as critical, as those in a cake recipe. A good cook won't use 5 pounds of flour if only 1 pound is needed just to get a larger cake—not unless he increases all the other ingredients proportionately.

WHAT IS CONCRETE?

Cement and concrete are not synonyms. Concrete results when specific amounts of various ingredients are mixed with clean water and then placed and allowed

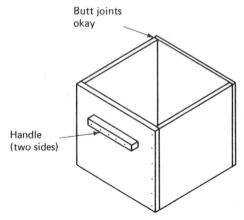

Butt joints
okay

Handle
(two sides)

FIG. 2–1   A bottomless box with inside dimensions to equal 12″ x 12″ x 12″ can be used to measure one cubic foot of material. Use ¾″ outdoor plywood for the box and solid stock for the handles.

to set and harden. Cement is just one of the recipe materials: The others are water and aggregates such as sand, gravel, and crushed stone.

Portland cement, which, incidentally, is a type of material, not a brand, is usually gray in color but is available in white. The latter costs more and is used mostly for special effects. Portland cement is available in 1-cubic-foot (ft.³) bags that weigh 94 pounds (lb.) each. In a suitable state the cement must be free-flowing and free of lumps that can't be easily pulverized with your fingers. Harder lumps are a good indication that the material has absorbed some moisture, which makes it less than suitable for critical work. When the moisture

FIG. 2–2   A contractor's wheelbarrow is a good, readymade trough for mixing batches of concrete. When working by hand, mix all materials thoroughly in dry state before adding the water.

FIG. 2–3   Correctly proportioned and mixed concrete ends as a workable mass. Light troweling will fill all the spaces between the pieces of aggregate with the paste formed by the cement and water.

content is not excessive, the material may be passed through an ordinary household screen and used for more detailed minor work. It is not likely that newly purchased cement will be lumpy, but, in storage, a bag or partial bag may absorb moisture. Always place cement in a dry area, raised above the ground on a simple wooden platform, and protected from the weather.

Don't confuse moisture lumps with what the trade calls "warehouse pack." The latter condition appears to be a hardening around the edges of the bag but is easily eliminated by rolling the bag on the floor. To minimize warehouse pack, suppliers are advised not to stack bags more than seven high.

When combined with water, the cement should make a paste that surrounds and binds together all the individual pieces of aggregate in the mix.

AGGREGATES

Aggregates are either "fine" or "coarse" sand and gravel (or crushed stone). The most common fine aggregate is natural sand but in some areas it is manufactured by crushing stones or gravel. The important factor is the size of the sand particles. For concrete, they can be as large as ¼ inch (in.) and range down to dust-size. The sand that is supplied especially for mortar (brick and block work) is not recommended for concrete, since it is composed of small particles only.

FIG. 2–4   In this mix, the amounts of aggregate are too little for the cement paste that is present. The project would lack durability and have a strong tendency to crack. Essentially, the mix is too wet.

FIG. 2–5   When a mix contains excessive amounts of fine and coarse aggregates it is very difficult to place correctly and to finish properly. Such a mix is designated as being too
stiff.

Coarse aggregate can range from the maximum sand size (¼ in.) to 1½ in.
Economically, a concrete mix with the largest, practical, maximum size of coarse
aggregate is best. Generally, the largest aggregate size should equal about one
third the thickness of the slab. For example, specify a 1-in. maximum size for
slabs that are 4 in. thick; 1½-in. maximum size for slabs that are 5 to 6 in.
thick. For such projects as concrete steps, where the thickness of the pour might
vary, specify a 1-in.-maximum-sized aggregate.

There is some leeway in size selection, so when you can't get exactly what
you want, go along with the closest, largest size available.

All aggregates should be clean. Excessive foreign materials such as dirt,
leaves, and common stones can result in a low-quality concrete because they
prevent the cement from doing a proper binding job. When possible, store
aggregates on a clean, hard surface. If you must dump on soil, it will pay to
cover the ground with a sheet of heavy plastic film. Also, it is a good idea to
cover the pile so that it will not become wet should it rain.

Whereas the big pieces of aggregate provide most of the bulk in a concrete
mix, the smaller pieces fill in the empty spaces. That is why a good aggregate
mix will contain all aggregate sizes without too much of any one. To visualize
why this should be so, picture a large box filled with basketballs and compare it
to a similar box that contains basketballs, baseballs, Ping-Pong balls, and marbles. You can see that a cement paste will have little difficulty filling the voids
in the second box but that the first box will result in a weak, porous product.

FIG. 2–6   When a mix contains too much fine aggregate (sand) and less than the called for
amount of coarse aggregate it is called too sandy. Placing and finishing would go
okay but the project might crack.

FIG. 2–7   Here, the mix is too stony. The coarse aggregate amounts have been overdone and
there is too little sand. This type of mix would be tough to place and finish; the
project would be weak and porous.

WATER QUALITY

An easy-to-apply standard for the water that you use in a concrete mix is that
the water should be fit to drink. This automatically eliminates water that contains impurities and foreign matter.

THE ROLE OF AIR

The important part that air can play in a concrete mix is a fairly recent dis-
covery. The presence of evenly dispersed microscopic bubbles of air in concrete
can contribute substantially to durability and to the minimizing of scaling that
often results from freeze–thaw actions.

Consider that most hardened concrete has some water content. As it freezes,
the expanding water can cause enough pressure to rupture the surface of the
concrete. When tiny bubbles of air are present, they provide relief areas for the
pressure and so prevent damage. Air bubbles also increase the workability of a
mix, so that, often, less water is required.

The procedure used to introduce air into concrete is called *air entrainment*.
It is a great asset in freeze–thaw situations. Even in comparatively mild climates,
where freeze–thaw cycles are minimal, air entrainment is advisable for all exterior concrete work.

The procedure involves the adding of an air-entraining agent to the mixing
water. Since the amounts required will vary from brand to brand, be sure to
read all instructional literature or consult your supplier. It is also possible to
buy portland cement with an air-entraining agent already ground in. Such products will be identified on the package as "air-entrained" and should be available
from any supply house that sells the regular portland cement.

A GOOD MIX

Quality concrete begins by measuring the ingredients carefully. When doing your own mixing, the most accurate way is to go by weight. But the individ-

ual, novice or pro, is not likely to have a proper scale on hand, so the next best method is to work by volume. Here, though, you can achieve good accuracy by proper use of a good container, such as a 3- or 5-gallon (gal) galvanized pail or bucket. Trying to judge amounts by the shovelful is *not* a good approach. One possibility when working by weight is to use a pail and a bathroom scale, checking the weight of a full pail, a half pail, and a quarter pail of each ingredient and marking the pail accordingly. Other batches won't have to be weighed since you will have the marks on the pail to go by. Be aware, though, that the moisture content of the materials can change and this will affect weight; so be sure to run a frequent check to maintain the accuracy of the marks on the pail.

Another accurate way, by volume, is to make a bottomless box with inside dimensions that equal 12 by 12 by 12 in. (1 ft³). This can be marked in halves and quarters and used by placing on a level surface and filling to the right mark with a shovel. With contents in place, simply lift the box straight up and then transfer the materials to the mix area. The box may be used in a wheelbarrow or a special mixing trough. Just be sure that the box is level, even when it is sitting on ingredients already measured.

MOISTURE IN THE SAND

It will be a rare day when bulk sand is delivered dry. The moisture it contains must be considered part of the mixing water. That is why the chart shown later lists sand in various degrees of wetness and suggests simple methods for testing by hand. In all cases, the wetter the sand, the lesser the amount of mix water. Generally, especially when the sand's moisture content is in doubt, be frugal with the water; you can always add a bit more.

AMOUNT OF WATER

Any person who has done any amount of concrete work knows that the wetter the mix is, the easier it is to work. But he should also be aware of the problems —from structural failure to chipping and flaking—that are the rewards of taking the easy way out.

Cement reacts with water to form a paste which hardens and bonds together all fine and coarse aggregates in a solid mass that is strong and durable. Some water is essential to trigger the chemical reaction that transforms loose, dry ingredients into "stone," but in almost all cases stronger concrete results when a minimum amount of water is used.

Water content should be sufficient for the hydration of the cement. Excess amounts occupy space in the mix and so displace other necessary ingredients. The captured water will eventually evaporate and leave voids that weaken the concrete. The voids may appear even on surfaces. Also, especially in a slab pour, excess water can float much of the cement to the surface, where it simply flows off and is wasted.

To avoid these and other problems, follow closely the amounts of mixing water called for in the charts.

TYPES OF MIXES

There are a number of ways to buy the ingredients for a concrete mix:

Buy all the materials dry, separately, and proportion the amounts yourself. This would include all the fine and coarse aggregates and the cement.

Buy a dry mix with aggregate sizes mixed to your specifications. To this you add the correct amount of cement and water.

Buy a dry ready-mix. Such mixes are available in sacks of various sizes, the most common being 45 lb. and 90 lb. A 90-lb. sack should produce about $\frac{2}{3}$ ft.^{3} of concrete. With these, you add the water and do the mixing.

Buy wet ready-mix that comes in a large, special truck and is used immediately upon delivery.

Your choice as to what to buy will be affected by various factors—the amount of mixing and proportioning you wish to do, comparative costs, the size of the job, the overall scope of the projects you plan. The least demanding method, in terms of prepour work, is wet delivery. Here, your only concern is the dumping, spreading, and finishing of the material. Also, when you deal with a reputable supplier, you know that the mix contents are technically correct. It is not likely, though, that such deliveries will be made for less than 1 cubic yard (yd^3).

Dry, bulk systems are good for large or small jobs since you can decide what amounts to mix. This is probably the most economical method, but it does involve more work on your part. This method is particularly useful when you have a series of separate projects. You will have a private source of supply at your disposal for use when the mood is right. Bear in mind that the materials, especially the cement, must be correctly stored to avoid contamination and absorption of moisture.

The dry-sack method is very convenient since the product is easy to store safely or to buy as needed. For large jobs, compare the cost of sacks and bulk. Regardless of cost, however, the sack method is fine for comparatively small jobs, such as setting fence posts, building stepping stones, making repairs, or for a crafts project.

Sack materials are subject to the same considerations as bulk materials, especially in freeze–thaw situations. Here, an air-entraining cement must be used, or an air-entraining agent must be added to the mixing water.

Be sure to read all the instructions on the package. They will tell you specifically what the contents may be used for, how much water to add, and how to mix correctly.

HAND MIXING

Hand mixing is acceptable for many types of jobs, although it is not recommended when air entrainment is involved because it is impossible to mix vigorously enough to achieve good results. For other work, however, you can mix in a wheelbarrow, a special trough made for the purpose, or on a hard, clean surface. A smooth concrete slab or a panel of tempered hardboard make good working surfaces. A good general rule is to thoroughly mix all materials in the dry state before adding water.

Start by spreading out evenly the correct amount of sand. Add the cement by distributing it uniformly over the sand and then working the two ingredients with a square-ended shovel until you achieve a uniform color. There should be no prominent areas or streaks of a single color. After this, even out the material and add the correct amount of coarse aggregates as a top layer. Again, work with the square-ended shovel until you are satisfied that you have a uniform blend of all the dry ingredients.

So that water won't be lost because of runoff, form a hollow in the center of the pile and slowly add the correct amount of water. Work from the perimeter of the pile to move dry materials to the water. Continue mixing until all the dry materials and the water have been thoroughly combined.

This is a good procedure to follow regardless of whether you mix in a wheelbarrow, in a trough, or on a flat surface. Do remember that, ideally, the paste formed by the cement–water combination should coat each piece of aggregate in the mix. Place the concrete as soon as the mixing is accomplished.

MACHINE MIXING

A fine way to proceed, especially when you have more than a few feet of concrete to lay, is to mix by machine. Mixing machines vary in size up to about 6 ft^3 and may be purchased or rented. For a single job or for projects with a time lag between them, it is wise to think of renting. For extensive work, buying a mixer may be an economical move; it will certainly save physical effort and contribute to a competent job. Finally, mechanical mixing is a must when air entrainment is involved.

The portable mixer is powered either by gasoline or electricity. The advantage of the gasoline model is that it can be used anywhere. When using such a machine, be aware of its maximum batch size (usually stated on an attached identification plate), and never overload it. Do not switch on the mixer until all the coarse aggregates and about 50 percent of the mixing water, plus the air-entraining agent (if any), are in the drum. With the mixer going, add the remaining ingredients—sand, cement, and the balance of water. Keep the mixer going until the contents have achieved a uniform color. About three minutes or so in a mechanical mixer should accomplish this.

Concrete should be placed as soon as possible after the mixing is accomplished. It is difficult to establish specific time limits, but, as a guide, consider that concrete that is not placed within an hour or so and that shows signs of stiffening may be remixed for a few minutes to restore workability. This doesn't always work, however. If, after the remixing, the concrete is still too stiff to be workable, discard it. Never attempt to use additional water as a means of restoring workability.

QUESTIONS

2–1	State a prime reason for bad concrete.
2–2	Tell the difference between *cement* and *concrete*.
2–3	What is the weight of a cubic-foot bag of cement?
2–4	Describe the job that cement does.
2–5	Describe aggregates.
2–6	State a general rule concerning the largest aggregate size that a mix should contain.
2–7	Describe why different sizes of aggregate are important.
2–8	Describe the quality of water that should be used in a mix.
2–9	What is *air entrainment?*
2–10	Why is air entrainment important?
2–11	Describe a box that will accurately measure ingredients by volume.
2–12	What can result when you use too much water in the mix?
2–13	Describe the various ways of buying mixes.
2–14	Describe a procedure that is acceptable when mixing by hand.
2–15	When should concrete be placed?

3

Some Basic Facts

A yard of concrete is quite a bit of material. To picture it, visualize a cube that measures 3 × 3 × 3 ft. Slice it into thirds from three directions and you get 27 cubes, each measuring 1 × 1 × 1 ft. If you make cuts 4 in. apart from one side, you end up with nine pieces, each measuring 4 in. × 3 ft × 3 ft or, 9 ft² of concrete 4 in. thick. The total of the nine slabs would equal 81 ft².

To estimate the amount of concrete required for any job, just multiply the thickness of the project by the width and that number by the length. Divide the result by 12 to find the number of cubic feet. The formula is

$$\frac{T \ (\text{in.}) \ \times \ W \ (\text{ft}) \ \times \ L \ (\text{ft})}{12} = \text{ft}^3$$

Thus a walk that is 4 ft wide × 24 ft long and 4 in. thick would require 32 ft³ of concrete; a wall that is 5 ft high × 20 ft long and 10 in. thick would require 83.33 (or 83.34) ft³. The formula and the charts do not include a waste allowance for unrecoverable spillage and for an uneven subgrade. To be safe, add a minimum of 5 percent (10 percent is better) to the estimate. Too much is a great deal better than too little; the excess, if any, can be used for a couple of small projects such as putting in stepping stones.

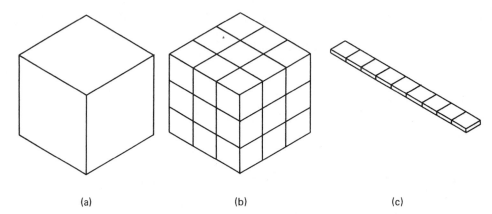

(a)   (b)   (c)

FIG. 3–1   How to see a yard of concrete: (a) a cube that measures 3′ x 3′ x 3′; (b) 27 cubic feet; or (c) nine slabs 4″ thick and each measuring 4″ x 36″ x 36″. Assuming no spillage or waste, the slabs would make a walk 3′ wide x 27′ long.

WHEN TO DO YOUR OWN MIXING

Small projects that involve only a few cubic feet of concrete are not difficult to do on your own. Sometimes, large projects can be reduced to a series of comparatively smaller efforts through formwork design—for example, using permanent grids in a patio. Here, though the total job may measure hundreds of square feet, it can be done in separate pours each of which covers, say, about 25 ft². This serves to make the job more feasible, especially for a beginner, and eliminates the awe of facing a large amount of material that seems to defy your strength, skill, or speed.

When the material requirements are beyond the capacity of your mixing container, it makes sense to have a helper. Sharing the chores with other workers can bolster your morale and contributes to a better result.

MATERIALS NEEDED TO MAKE ONE CUBIC FOOT OF CONCRETE WHEN PROPORTIONED BY VOLUME

Coarse Aggregate (max. size, in.)	Air-Entrained Concrete*				Concrete Without Air			
	Cement	Sand	Coarse Aggregate	Water	Cement	Sand	Coarse Aggregate	Water
⅜	1	2¼	1½	½	1	2½	1½	½
½	1	2¼	2	½	1	2½	2	½
¾	1	2¼	2½	½	1	2½	2½	½
1	1	2¼	2¾	½	1	2½	2¾	½
1½	1	2¼	3	½	1	2½	3	½

*See text for explanation of air-entrained concrete.

MATERIALS NEEDED TO MAKE ONE CUBIC FOOT OF CONCRETE WHEN PROPORTIONED BY WEIGHT

Coarse Aggregate (max. size, in.)	Air-Entrained Concrete* (lb.)				Concrete Without Air (lb.)			
	Cement	Sand	Coarse Aggregate†	Water	Cement	Sand	Coarse Aggregate†	Water
3/8	29	53	46	10	29	59	46	11
1/2	27	46	55	10	27	53	55	11
3/4	25	42	65	10	25	47	65	10
1	24	39	70	9	24	45	70	10
1 1/2	23	38	75	9	23	43	75	9

*See text for explanation of air-entrained concrete.
†For a crushed stone mix, *decrease* the amount in the coarse aggregate column by 3 lb. and *increase* the amount in the sand column by the same amount.

MATERIALS NEEDED FOR SPECIFIC BATCHES OF CONCRETE*

Concrete Required		Cement† (lb.)	Water (gal.)	Sand (lb.)	Coarse Aggregate (lb.)
Cubic Feet	Cubic Yards				
1		24	1 1/4	52	78
3		71	3 3/4	156	233
5		118	6 1/4	260	389
6 3/4	1/4	165	8	350	525
13 1/2	1/2	294	16	700	1,050
27	1	588	32	1,400	2,100

*The proportions are for a mix consisting of 1 part of cement to 2 1/4 parts of sand to 3 parts coarse aggregate. Aggregates for this mix should be no larger than 1 in. in diameter.
†In the United States, a bag of cement weighs 94 lb; in Canada, 80 lb.

MATERIALS NEEDED FOR SLABS OF CONCRETE*

Thickness of Slab (in.)	Project Width Multiplied by Length† (ft.²)					
	10	25	50	100	200	300
4	0.12	0.31	0.62	1.23	2.47	3.70
5	0.15	0.39	0.77	1.54	3.09	4.63
6	0.19	0.46	0.93	1.85	3.70	5.56

*Assumes a perfect subgrade (level) and does not allow for waste and spillage. To be safe, add 5–10 percent to the estimate.
†For projects that are larger than the examples in the chart, choose the columns that add up to the size you require. For example, for a project that contains 550 ft², add the figures in the 300, 200, and 50 columns.

Never neglect the site preparation and formwork that precedes the pour. Extra time spent on this phase of the operation assures smoother pouring and finishing. Few things are as disastrous as forms that collapse or bow out, stakes that don't hold sideboards in place, grids that don't stay straight, and the like. Minor corrections can be made, but wet concrete can be intolerant of major ones.

When mix requirements approach 1 yd, consider having the wet, ready-mixed materials delivered to the site by truck. Most people fear doing this because of a false view of what will happen. Be assured that the driver is not about to dump the contents of the truck in a pile on your doorstep and then take off. He is aware that the concrete must be placed and is prepared to spend a reasonable amount of time to do it correctly.

His cooperation is guaranteed if the site and the forms are ready and if you have visualized and prepared for what must be done once the pouring starts. Conquer the fear of wet delivery and you will get to using the method even when a full yard (if that is the minimum delivery amount) is not required. The excess can be used for smaller projects that you have ready to go. These can include stepping stones, landing ramps, downspout runoff troughs, small curbs, even a craft project.

SUGGESTED MIXES FOR TYPES OF PROJECTS

Gallons of Water per Sack When the Sand is*:

The Project	Dry	Damp	Wet	Very Wet	The Mix	Coarse Aggregate (max. size, in.)†
Subject to extreme wear and weather: heavy-duty	5	4¼	4	3½	1–2–2¼	¾
General slabs walks, patios, floors	6	5½	5	4½	1–2¼–3	1
Foundations footings, walls	7	6¼	5½	4¾	1–3–4	1½

*Dry sand lacks moisture; damp sand will not compact if you squeeze a handful; wet sand compacts when you squeeze a handful, but it will not leave the hand wet; very wet sand will leave considerable moisture on your hand.
†A general guide is to restrict the maximum aggregate size to about one third of the slab thickness.

WORKING IN COLD WEATHER

If you have a choice, do your concrete work during temperate weather conditions. If you must work during cold weather, be aware of the following facts.

Do not place concrete on frozen ground. Straw mats, loose straw coverings, and similar materials can be used to keep the earth from freezing before the concrete is poured. Check over the site, the forms, and any reinforcement materials (bars or mesh) and remove any accumulations of ice or frost.

Correct hardening of the concrete will occur when the temperature of the mix falls between 50° and 70° F. If the temperature falls below 50°, you can heat the mixing water, the sand and gravel, or both. It is possible to get heated water by running a garden hose from the kitchen or from a utility sink in the garage. Short of this, you can heat water by suspending buckets over outdoor, open fires.

Aggregates can be warmed with hot water, or you can place them on a raised, metal platform under which there is room for a fire. Frequent folding with a shovel will help assure uniform heating. The aggregates should feel hot to the touch but they should not be burned. In some areas, it is not uncommon for suppliers of wet, ready-mix to make it available in heated form.

Once the warmed concrete has been placed and finished, you must continue to protect it from freezing. Usually, coverage with a tarpaulin and a thick layer of straw will do the job. Supplementary heat, supplied by space heaters, has been used successfully in areas where the temperature is consistently low.

WORKING IN HOT WEATHER

The problems posed by hot weather are more easily solved. Just use water to keep the temperature of the concrete between 70° and 90° F, and to cool down forms and subbase materials. Place and finish the concrete as rapidly as possible.

CURING

The chemical reaction between water and cement (hydration) is responsible for the tremendous bond that grips concrete aggregates together. Hydration works slowly and is in effect only as long as water is present. If the water evaporates, hydration stops. In general, the good qualities of concrete will be minimized and the material will cease to gain strength. In essence, curing means assuring that little or no moisture will be lost from the concrete, especially during the initial hardening stages. The truth is, that even when the concrete is finished and surface-hardened so that it will resist abrasion, you could submerge it in water and the act would do nothing but good.

Curing can make or break a concrete project, regardless of the energy and dedication expended up to that point. Since it is such a simple operation it doesn't make much sense to neglect it.

A technique used often, especially by the do-it-yourselfer (since he is usually physically close enough to the job to do it), is to wet the project with a fine spray from a garden hose at very frequent intervals. The frequency of spraying can be minimized if the concrete is covered with a moisture-absorbing material such as burlap. Newspapers, soaked in water, are often used, but the possibility exists that the printer's ink may be picked up by the concrete.

Special kraft curing papers, even plastic sheeting, can be used as covers and they do not require wetting. Such materials will keep existing water from evaporating. It is a good idea, however, to wet them down anyway; it will certainly do no harm.

Special, commercial compounds are available for curing. These are sprayed on the surface immediately after the concrete is finished to form an antievaporation membrane.

Curing standards call for the process to continue for a minimum of 72 hours —six days should the temperature drop to 50° or less. If you have concrete work done by a contractor, it is normal for him to start the curing process but for you to continue with it. Usually, this doesn't involve more than keeping the original covering in place and wet.

In any situation, it doesn't hurt to keep the curing process going for at least a week.

FIG. 3–2   This concrete project is being wet-cured with a covering of burlap. The material holds moisture longer so its use minimizes the frequency of water applications.

QUESTIONS

3–1   Describe a yard of concrete.
3–2   How many square feet of concrete would result if you were to use a yard of concrete (totally) for a slab that is 4 in. thick?
3–3   State a formula that can be used to estimate the quantity requirements for a concrete job.
3–4   State some of the facts you should be aware of when you do concrete work in cold weather.
3–5   What is curing?
3–6   Describe several methods of curing concrete.
3–7   How long should you cure?
3–8   Describe how you can judge the moisture content of sand by squeezing it in your hand.

4

Tools for Concrete Work

A pleasant aspect to concrete work is that much of it can be done either with tools that you already own or with tools that you can make. If you do have to buy, you'll find that, with rare exceptions, most items fall in the price range of good garden tools. Of course, you can save some money initially by buying bargain-counter equipment, but you'll lose out in the long run. The rule to follow is to buy topnotch tools and to take care of them.

TOOLS FOR SITE PREPARATION

A *mattock*, often called a "pickax," is used for such jobs as loosening soil that is to be removed and for digging trenches. It has a sharp point at one end of the head and a chisel-like edge at the other. The broad edge is very useful for working on a horizontal plane when you wish to remove a top layer of dirt with minimum disturbance of the soil beneath.

A *short-handled shovel* with a square edge, often called a "square-point," is a much better tool for concrete work than the garden variety of long-handled, round-pointed shovel.

You will need a *tamper* for firming loose soil and for compacting subbase materials. Commercial varieties are long-handled and have a heavy, steel pad. If you prefer, you can make wooden ones that will do acceptable work.

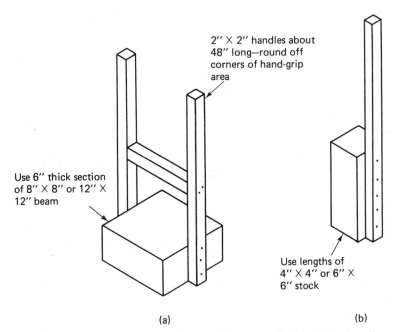

2" X 2" handles about 48" long—round off corners of hand-grip area

Use 6" thick section of 8" X 8" or 12" X 12" beam

Use lengths of 4" X 4" or 6" X 6" stock

(a)   (b)

FIG. 4–1   How to make a hand tamper: (a) double-handle type and (b) single-handle type—good for small jobs and for getting into corners.

A rake for leveling soil or dumped materials can be of the garden type but should be made of steel and strongly constructed.

TOOLS FOR FORMWORK

For crosscutting and ripping, you will need an 8-point, 24-in. utility saw. A 10-point, taper-ground, crosscut saw will leave smoother edges and is a particularly good tool to consider for permanent forms—for example, grid pieces in a patio or walk.

FIG. 4–2   Screeding is done after the concrete is poured and spread. The tool may be a straight piece of 2 x 4 or 2 x 6. It rests on the side forms and is moved in zigzag fashion as it is pulled toward the worker.

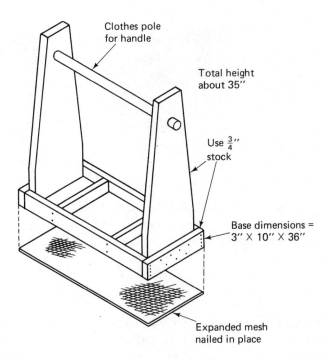

Clothes pole
for handle

Total height
about 35"

Use $\frac{3}{4}$"
stock

Base dimensions =
3" × 10" × 36"

Expanded mesh
nailed in place

Fig. 4–3   How to make a jitterbug tamper. The jitterbug tamper is used after screeding to
prepare the surface for floating operations. Don't overdo the application or you may
force too much of the coarse aggregate to the bottom of the pour.

A 20-ounce (oz) *framing hammer* is a better choice than the common
16-oz cabinetmaker's tool, because it has more heft for driving large nails. In addition, you will want an 8-lb double-faced *sledge hammer* for driving stakes. A
handle about 32 in. long will facilitate heavy-duty work. This tool may also be
used for breaking up concrete or rocks. Be sure to wear safety goggles on any
job where the impact might throw off chips or debris.

A *half hatchet* is a good utility tool. The broad, sharp edge is used for
pointing stakes, the hammerhead end for driving nails and small stakes.

A *line, line level,* and *flexible tape* are all highly useful. Lines are available
in 200-ft rolls and are either white or yellow so that they are easy to see. Use
them to mark off perimeters of job sites and lines of forms and grid patterns.
The level is a small tool designed so that it may be attached at any point on a
stretched line. Flexible tapes are available in various lengths. Consider getting
an 8- or 10-ft size for short measuring and a longer one (at least 25 ft) for overall layout and checking.

TOOLS FOR MIXING

The same *shovel* described under "Site Preparation" is usable here. A *hoe* can
also be a help. The garden type is often used, but a better choice is one de-

Fig. 4–4   Low-handled darby is used much like an oversized float. Such tools are not difficult
to make. Professionals will often design one to suit a particular application or personal preference.

signed for the purpose which has a couple of large holes through the blade. The
holes make it easier to push and pull the tool through the ingredients and, to
some extent, aid the mixing process.

TOOLS FOR SPREADING

Use the *shovel* to move mixed concrete about. The square edge of the tool may
be used in limited fashion to get concrete into corners; a *hoe* or *rake* may also
be used. A garden rake works okay, or you can use a special concrete rake that
is double-bladed. One edge has teeth, the other is straight: The serrated edge
is used for spreading, the other for leveling and preliminary floating.

TOOLS FOR FINISHING

Screeds or *strikeboards* are used to level the concrete to the height of the forms.
They are moved in zigzag fashion to level the pour and to push excess concrete
toward a border. Often, even pros will employ a length of straight 2 × 4 for the
purpose. Commercial types, which last a lifetime, are made of lightweight metal
and come in various sizes and lengths.

 Tampers used after the concrete is leveled are mostly of the *jitterbug* variety. Short, quick, up-and-down strokes settle coarse aggregate and bring a finishable amount of cement paste to the surface. This phase of the job should
never be overdone, or too much of the coarse aggregate will be pushed to the
bottom of the pour. One of the drawings in this chapter will show you how to
make a jitterbug tamper.

 After tamping, actual finishing begins by using a wooden *float*. Floats are
available commercially in various sizes but are also easy to make. A float is used
with overlapping arc strokes to create surface uniformity. Quite often, this is
the only finishing technique used; the result is a finish that is reasonably smooth
but that has good traction.

 Bull floats are often used to achieve the same results obtained with hand

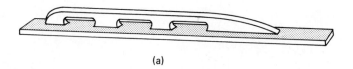

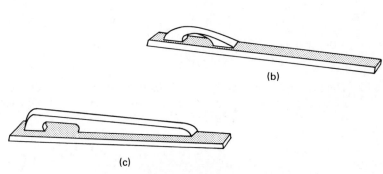

FIG. 4–5   Types of darbies: (a) 45″ long, three-grip handle, ¾″ thick base, tapers from 3⅝″ at the rear to 2½″ at the point; (b) single-grip magnesium is light-weight and easy to handle, 30″ long, tapers from 3½″ to 2¼″; and (c) single-grip wood is 28″ long, tapers from 3½″ to 2¼″.

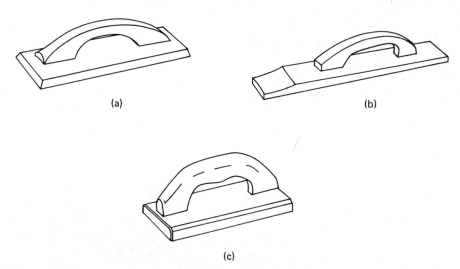

FIG. 4–6   Types of floats: (a) common *wide* float—two popular sizes are 4½″ x 12″ and 4½″ x 15″ (b) bevel-edge float—choice for close formwork, faces are beveled, two typical sizes are 3½″ x 15″; and 3½″ x 18″; and (c) molded rubber float—texture is very fine and even, rubber pads are replaceable, common sizes are 4″ x 8″ and 4″ x 10″, similar float has a permanently bonded cork surface.

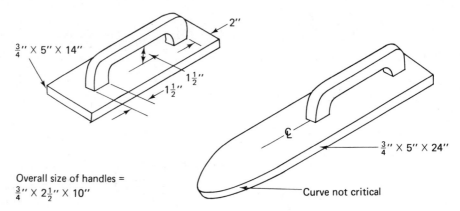

$\frac{3}{4}'' \times 5'' \times 14''$

2''

$1\frac{1}{2}''$

$1\frac{1}{2}''$

$\frac{3}{4}'' \times 5'' \times 24''$

Curve not critical

Overall size of handles =
$\frac{3}{4}'' \times 2\frac{1}{2}'' \times 10''$

Fig. 4–7   Here are two examples of wood floats that you can make yourself. They may be made of pine but will last longer if a hardwood is employed. Use screws and waterproof glue to attach the handles.

Fig. 4–8   Note the blade angle of the bull float. This is the correct position when pushing; the opposite is true when pulling. Keeping the blade flat would result in pushing the concrete instead of smoothing it.

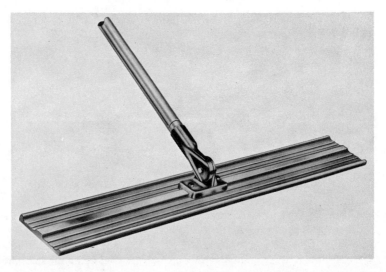

Fig. 4–9   Commercial, metal bull float has a pivoting handle that may be positioned anywhere in a 180° arc. Thus you can work from any side of, and to any point in a pour.

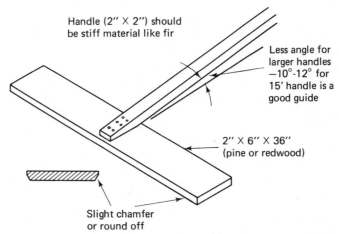

Handle (2″ × 2″) should
be stiff material like fir

Less angle for
larger handles
—10°-12° for
15′ handle is a
good guide

2″ × 6″ × 36″
(pine or redwood)

Slight chamfer
or round off

FIG. 4–10 It doesn't pay to buy a bull float unless you plan extensive or professional use. For around-the-house work, something along the design of the one shown here will work okay.

FIG. 4–11 Trowels are made of metal, are used pretty much like wood floats, but produce a slicker finish. This is not usually done on slabs that should have good traction for safe footing.

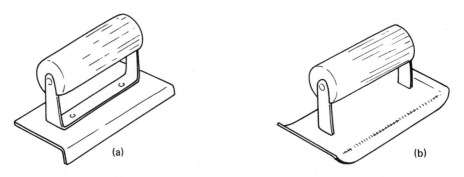

(a)   (b)

FIG. 4–12 Edgers (a) generally run from 6″ to 10″ long. Radius range is from ⅛″ to 1″. A new type edger (b) has curved ends and many find this style easier to use.

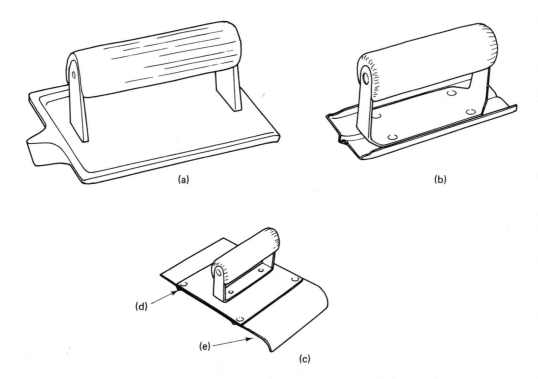

Fig. 4–13 Groovers: (a) Overall size averages about 4½″ wide x 6″ long. Width and depth of groove varies; (b) cheater cuts a groove that simulates expansion joint; and (c) combination type cuts groove (d) and forms edge (e). Useful, for example, when curb and sidewalk are cast in one pour.

floats. The former are larger and usually have extension handles so that areas of large pours may be reached. It is also true that a bull float may be used to prepare a concrete surface for finishing with a hand float.

Darbies, in essence, are oversized floats. They enable the worker to cover more area and to reach more places without having to walk on the fresh concrete.

The design of a *trowel* is similar to that of a float, but trowels are always made of metal. A trowel is used once the float has done its job to get a slicker finish than you could obtain otherwise.

Edgers are special tools with a flange at right angles to the base. The joint between the flange and the base is a gentle arc, and this forms the concrete edge when the tool is used. The idea is to avoid a sharp edge that can easily chip or crack.

Groovers are used to form surface cuts in the concrete. The depth of the cut depends on the purpose of the groove—a problem we will discuss later. Grooves are often included as control joints; other times they are merely decorative. Groovers designed for the latter purpose are often called *cheaters*.

FIG. 4–14 Power troweler uses gasoline instead of muscle but, more important, it can get a large slab done before the concrete sets enough to be troublesome. Tools like this can be rented.

FIG. 4–15 Stiff or soft-bristle floor brooms may be used to achieve particular types of texturing. They may be used in straight, parallel strokes or in swirl fashion. Soft bristles will create less texture.

F<small>IG</small>. 4–16 The broom shown in these photos is a special exposed aggregate broom that includes water jets. We will talk more about it later in Chapter 8, Special Finishes.

CARE OF TOOLS

Never walk away from a job without cleaning the tools that you have used. Water-cement paste, for example, will cling to a tool, harden quickly, and become difficult to remove. The best way to remove excess material from a tool is to scrape it with a small piece of softwood. Remove what remains by dipping the tool in water and rubbing it with a cloth or brush, or by blasting it with a garden hose. Wipe the tool dry with a cloth. If you plan to store a tool for any length of time, protect it with a very light film of oil. The oil should be removed before using the tool again.

QUESTIONS

4–1   List the tools required for site preparation.
4–2   List the tools required for form work.
4–3   List the tools required for hand mixing.
4–4   List the tools required for spreading concrete.
4–5   List all the tools that are used after the concrete has been spread. Describe briefly the use of each.
4–6   Describe how tools should be cared for.

5

Preparing the Subgrade

Concrete failures can be caused by a poorly prepared subgrade. This part of the job is often neglected, especially by the beginner, who is anxious to get on to the forming and pouring. The results of slighting this part of the process can be discouraging. Slab settlement—wholly or in part, large cracks, outright structural failure, all can occur. Serious attention must thus be paid to preparing a subgrade that is hard enough to take the weight of the concrete. It should also be uniform, free of foreign materials such as rocks, roots, and grass, and it should be well drained.

Undisturbed soil makes a pretty good subgrade, so when you must remove a uniform amount of dirt, do the digging so as to loosen only the soil that will be removed. Fill in existing holes, or holes caused by removal of loose soil, with sand or gravel and compact thoroughly. It is all right to do some refilling with soil that is similar to what is in the subgrade, but do it in layers of 3 to 4 in., each layer well compacted and evened out. Be sure that fill materials do not contain big stones, lumps, or foreign matter.

Granular fills (crushed stone, sand, gravel) are highly recommended for bringing the site to final grade and for providing uniform support for the concrete. These materials, though, should not be merely shoveled into place and raked. It is much better to work in layers of 3 or 4 in. with solid compacting at each level. A thorough worker will extend the fill 1 ft or so beyond the edges of a slab so as to prevent the removal of subconcrete materials (undercutting) by rainfall.

Subgrades that drain poorly will be water-soaked most of the time and thus require special attention. As much as 6 in. of well-compacted granular fill should be used in such situations, and, to be more than safe, design so that the bottom of the fill will just abut finished grades.

It takes nature time to compact soil, so it is taboo to place concrete over any area that has been freshly filled unless specific precautions are taken. These usually consist of sinking concrete piers down to the original grade and then compacting the surface by machine. When the fill is slight, the pier holes can be dug with a shovel. Some craftsmen will drive a pipe or a preservative-treated length of 2 × 4 redwood down the center of the hole with its top edge on a level with the subgrade. When in doubt about what to do in a similar situation, it is wise to consult with the local building inspector, who should be able to tell you exactly how you should design. Codes vary geographically. In some areas of the country you can put down, for example, a 4-in. patio slab directly on firm soil. In other places, a good job may call for a certain number of inches of granular subbase materials. So it pays to check.

All subgrade and subbase materials should be compacted. On small jobs, this can be done by hand with a steel tamper that you can rent or with a heavy wooden one that you can make. For extensive work, big patios or long walks and driveways, it pays to rent a small mechanical roller or a vibratory compactor. These are professional tools but they can be used efficiently by the beginner. The rental agency should supply complete instructions; they may even put you through a short test run in their own backyard.

Be sure that the prepared site is uniformly moist just before the pour. Areas that are too wet are just as bad as areas that contain no moisture at all. Make

Fig. 5–1   Small vibratory compactors are good tools to consider when the job seems too big to be done with a hand tool. These are powered by gasoline engines and are completely portable.

Fig. 5–2   Mechanical rollers will do large areas efficiently and in little time. Rental agencies
            may have some units that are even smaller than this example, but practical for drive-
            ways, walks, etc.

certain that there are no pools of water, soft spots, or muddy spots. Otherwise, dry areas will suck more moisture from the concrete than wet areas and the result can be a splotchy concrete finish. Also, removal of moisture from the mix may hinder curing.

The wetting you do can be accomplished with a fine spray from a garden hose. Form boards, too, should be wet down.

QUESTIONS

5–1   Describe briefly a well-prepared subgrade.
5–2   What materials are recommended for bringing the site to final grade?
5–3   Describe the best way to install fill material.
5–4   What are the minimum precautions that should be taken when concrete is placed over fresh soil fill?
5–5   Why is it important to check with the local inspector before designing a subgrade?

6

Formwork for Concrete

The amount of formwork you must do depends on the job and, to some degree, the design of the project. A driveway or walk with nothing but control joints, for example, doesn't need more than side forms. A gridded patio requires the permanent intersecting pieces as well as the perimeter forms. Formwork for a wall calls for more attention than does a ground slab. In all cases, the wooden forms must be strong enough in themselves and must be braced adequately enough to withstand considerable pressure.

Rigidity and bracing are guards against collapse and uneven lines, and they prevent bulging and curves where you don't want them. It is easy to be impatient with this preliminary aspect of the job. Because forms are usually temporary, they are often viewed as a necessary nuisance. True or not, like a well-laid subgrade, good formwork leads to successful projects. This does not imply that a cabinetmaker is needed to do the work. Fancy joinery is not the thing here; well-fitted, tight butt joints are.

MATERIALS

Various types and sizes of materials are used: 2 × 4's are good for projects that must hold up under traffic; 2 × 6's for jobs designed for vehicular loads. Although there is some leeway and you will see exceptions to the rule, use 1 × 2 stock to make stakes for 2 × 4's and 2 × 4 stock to make stakes for 2 × 6's. 2 × 4's are always used as bracing for projects like walls.

Formwork for wall-type projects is usually constructed of 1-in. lumber (nets about ¾ in. thick) or plywood. Ordinary plywood is often used but contact with the concrete doesn't do the material much good. A better choice, especially for repeated use, is a special concrete-form plywood which is designed for the purpose. This product, usually ⅝ or ¾ in. thick, is specially treated for durability and to produce a smooth finish on the pour. If you are in a position to make a choice between lumber and plywood, take the plywood. It is less trouble to assemble because you can work with big pieces and it will produce a smoother finish.

One-time use of wood materials to make concrete forms can mean a considerable cash outlay. That's why many contractors will choose a material that is suitable, after cleaning, for such additional uses as subflooring or sheathing.

You can create different effects or interesting textures merely through the selection of a particular material for the form. Rough (unplaned), resawed, or sculptured lumber or plywood will leave a woodgrain effect on the concrete. Wood that is not so sound and has surface blemishes is often used in place of sound lumber just because of the special effect it produces. To do the opposite, that is, to get a supersmooth finish, you can line the forms with plastic sheeting or a heavy kraft paper. In essence, although you cannot sculpt a mass of concrete as you can clay, a great deal of creativity is possible through uncommon forming and imaginative selection of forming materials.

Such innovations, however, should be restricted to special applications.

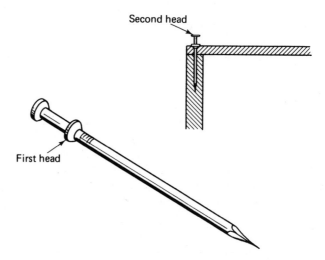

Second head

First head

FIG. 6–1   Double-headed or duplex nails are useful for temporary assemblies such as you need for concrete forms. The nail is driven to the first head. The second head projects so the nail is easy to remove.

There is not much point in getting a fancy finish on a wall that will be covered by backfilling. In this case, the only considerations should be strength, speed, and economy.

Professionals today use knock-down forms that can be used over and over again. These are very much like kits that can be assembled to suit various types of projects. In many areas of the country, such materials can be rented.

When assembling temporary forms yourself from raw materials, it is a good idea to use double-headed (duplex) nails for the fastening. Drive these nails to the first head to hold parts together, but leave a second head exposed for withdrawing. Such nails, of course, are not used for permanent formwork, for which you will work with regular, galvanized nails.

Wood forms will release from set concrete more easily and with less chance of surface damage to the concrete if they are treated with a "release" before the concrete is poured. An old standby is ordinary engine oil applied in a thin coat by brush or rags. There are many, more modern materials available that fall into the general category of "form releases." These are purchased in bulk and are applied by compressed air with a unit such as those used for spraying insecticide in the garden.

LAYOUT _____

The dimensions, the shape, and the location of the new project on the site must be carefully laid out before you can start formwork. This is not the kind of job you can "eyeball," nor can you be haphazard about observing restrictions that may be imposed by local codes. In most areas, you can build only so near property lines, and the open area required may differ from back to side to front. Working correctly now may save you the chore of a demolition job later. You will probably need a building permit anyway, for which there is a small fee; so you may as well get your money's worth from the people who issue it to you by getting all the information that they have available.

First, set out a base line to represent one side of the new project. Do this by measuring in from the property line or, if the project calls for it, from an existing building. Drive two stakes to mark the ends of the line and then set up batter boards (see drawings) outside those points.

To move away from the first line at a right angle, you can work in one of two ways. Either make the check gauge (large square that we show in a drawing) and use it just as you would a small carpenter's square, or work by actual measuring, as follows. Set up a temporary third stake approximately at right angles to one of the corner stakes but exactly 8 ft away. Tie a line to it and then cut it so that it is exactly 10 ft long. Now you can move the third stake laterally until the 10-ft line touches the base line at a point exactly 6-ft away from the corner stake. The corner so established will be a right angle. The system is merely the establishment of a right triangle that has an 8-ft leg, a 6-ft leg, and a

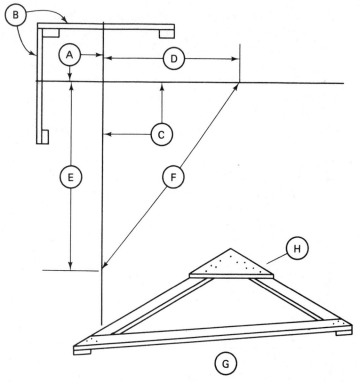

FIG. 6–2 Lines (a) stretched from batter boards (b) should form an inside angle of 90° (c). When you can measure 10′ at (f) from a 6′ point on one leg (d) and an 8′ point on the other leg (e), the angle is 90°. Another way is to make a check gauge (g). This has a 6′ leg and an 8′ leg and is reinforced at the square corner with a plywood gusset (h).

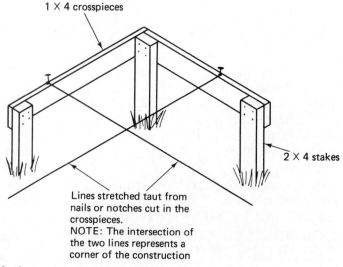

1 × 4 crosspieces

2 × 4 stakes

Lines stretched taut from nails or notches cut in the crosspieces.
NOTE: The intersection of the two lines represents a corner of the construction

FIG. 6–3 Basics of the batter board.

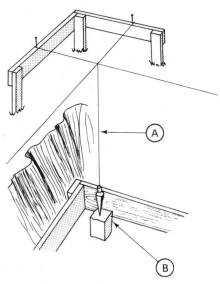

FIG. 6–4   How batter boards are used at an excavation: (a) drop a plumb bob from the line
intersection and (b) drive a corner stake.

10-ft hypotenuse. With two corners of the project established, one of them
square, you can stretch line to a third corner and set up batter boards and then
proceed to the fourth corner. When setting the final line positions, notch the
batter boards with a saw so that the lines may be removed when necessary and

FIG. 6–5   Working with stretched line, even when doing a gridded patio job, is better than
measuring spacing for individual pieces. It minimizes the possibility of human error.

returned to correct positions. As a final check, measure the two diagonals—they should be equal.

Be sure to stretch all lines taut and to secure them by wrapping and knotting them around a small nail that you drive into the batter boards. A plumb bob, dropped from the point where the lines intersect, will determine exact corner locations. In this case, you can drive stakes and tap in a small nail at the point determined with the plumb bob.

To establish the final grade, set one of the corner stakes so that it represents the height of the new project. From this point you can work with stretched line and a line level to determine the correct height of the remaining three stakes.

The preceding is a typical layout technique. To project from or into what you have established as a square or a rectangle, work from one of the perimeter lines and use the same method plus additional batter boards.

REINFORCEMENT: STEEL BARS

The two most common materials used to add strength to concrete are steel bars and wire mesh. Both should be clean and free of rust, scale, and other foreign coatings when used. The steel bars are ridged and are available in diameters from ¼ up to ¾-in. For average projects, ⅜- or ½-in. bars will do.

Fig. 6–6   Reinforcement rods are made of strong steel but are soft enough to bend and cut easily. Most popular sizes for average jobs are ⅜-inch and ½-inch but they are available as large as ¾-inch.

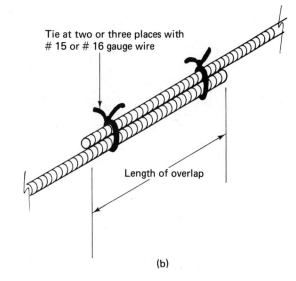

Diameter of the rod	Length of overlap
$\frac{1}{4}$"	12"
$\frac{3}{8}$"	18"
$\frac{1}{2}$"	24"
$\frac{5}{8}$"	30"
$\frac{3}{4}$"	36"

(a)   (b)

FIG. 6-7   Use one-piece reinforcement rods whenever possible. When the length of the project doesn't permit this, overlap the rods as shown.

Although the bars are fairly rigid and strong, they can be bent rather easily, especially in gentle curves and corners. Use an ordinary hacksaw to cut them by making a cut about halfway through and then bending sharply on the cut line to separate the pieces. Organize the placement of rods so that the sections around corners are a continuous, whole piece. When you must join rods for a continuous run, be sure to overlap the amount called out in Fig. 6-7. The amount of overlap depends on the diameter of the rod. Overlaps must be held

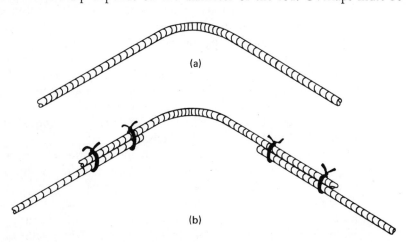

FIG. 6-8   Reinforcement rods around corners should be continuous (a) or lapped as shown (b) at one end or both ends. Be generous with both the overlap and the ties.

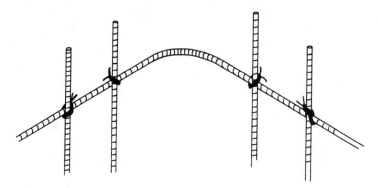

Fig. 6–9   Use wire ties wherever reinforcement rods cross.

together with two or three ties, depending on the length of the overlap, made from a No. 15 or No. 16 gauge wire.

The rods should be straight enough and placement true enough so that they will be amply covered by the concrete. If the rods are continuous down a wall and into a slab, for example, support the base portion of the rods on blobs of concrete so that the pour can flow under them. The idea is to elevate the reinforcement so that it will be approximately centered in the slab. Sometimes small blocks of wood or pieces of stone are used to do this. Remove such materials as the pour progresses.

REINFORCEMENT: WIRE MESH

The mesh is actually a welded wire fabric that is used most often in slab work, such as patios, driveways, indoor floors, and the like. The mesh is seldom used for patio work in mild climates, but when the soil is not stable or the project is to carry a structural load, using the mesh is a good idea.

When possible, place the mesh in whole pieces. If you must combine pieces to cover an area, overlap the joints about 6 or 8 in. or at least the width of one of the openings in the mesh design. As the material comes in rolls and has a tendency to curl, spread it over a flat area and walk on it to get it flat before you set it in place. Use heavy wire snips for cutting and work so that cut strands won't snap back against your body or face.

If you are forming up for both a slab and a footing or wall, use wire ties to hold the mesh to reinforcement rods. The least that this will do is keep the wire in place.

Wire mesh, like steel bars, should be suspended in the concrete in order to provide maximum strength. This can be accomplished by using small stones, pieces of brick, or blobs of concrete as "elevators." Many professionals will simply place the mesh flat on the ground and then raise it with a rake or a hooked

Fig. 6–10 Wire mesh is actually strong metal fabric with welded joints. It is used most often in slab work and it is easy to see how it works to increase the strength of the concrete.

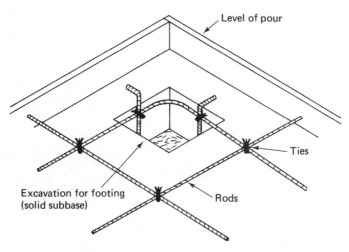

Fig. 6–11 When post footing is poured together with a slab.

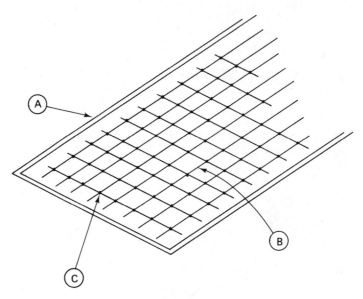

F<small>IG.</small> 6–12 Wire mesh used as reinforcement (mostly in slabs): (a) forms; (b) mesh between forms should be midway between surface of subbase and surface of pour; (c) may be raised *before* the pour on small pieces of brick or stone or blobs of concrete, or *during* the pour with a rake or hooked wire.

F<small>IG.</small> 6–13 Reinforcement rods in foundation work are often installed in such a way that ensuing concrete pours will tie into existing projects. Here, the bars will connect with a future patio pour.

wire as the pour progresses. Be sure not to raise the wire too high. A center position in the pour is what you should aim for, but the mesh should never be closer than about 1 in. to the surface of the slab.

PLASTIC UNDERLAY

Plastic underlay is available in 50- or 100-ft rolls and in 9- or 12-ft widths. It is placed over the subbase before the concrete is poured. Since its purpose is to prevent moisture from coming up through the slab, its use is generally confined to concrete floors.

The material is easy to cut and place but great care is required afterward to keep from piercing it. Be generous at perimeters; curl the plastic up and over forms. Any visible excess is easy to cut away after the concrete has set.

FORMS FOR SLABS

The final grade of a slab depends on the accuracy with which you set the side forms. Sometimes, especially for driveway and walk projects, it isn't necessary to set up batter boards. You can work by driving stakes at extreme points and then stretching a line between them. The top of one of the stakes should be established as the height of the project. Then you can work with a line and a line level to correct the height of the other. When the project is quite long, it will pay to set up intermediate stakes so as to avoid possible sag in the line. The line must tell project height and be a guide for setting the side forms in straight fashion.

The net size of lumber is less than the listed dimensions. Thus if you want a 4-in. slab, 2 × 4 or 1 × 4 side forms must be set slightly above the surface of the subbase. This also applies when you are using 1 × 6 or 2 × 6 material to get a 6-in. slab. If a gap is left between the bottom edge of the form and the top of the subbase, backfill with soil outside the forms so that the concrete will be retained.

All forms must be strongly held with stakes to keep them vertical and to prevent bulging. There is some leeway in the material that you can use for stakes; it may depend on what is available: 2 × 4's, 2 × 2's, 1 × 4's, 1 × 2's can be used. The heavier the stake, the fewer stakes you will need. As a general rule, when working with 2-in.-thick formboards, space stakes a maximum of 4 ft apart. When the formboards are thinner, use more stakes. In truth, if you are generous enough to supply stakes every 2 or 3 ft, you will be contributing a good safety factor to the project.

Be sure to drive stakes so that they are firm. Tie the form and stake together by driving double-headed nails through the stake and into the form. The top

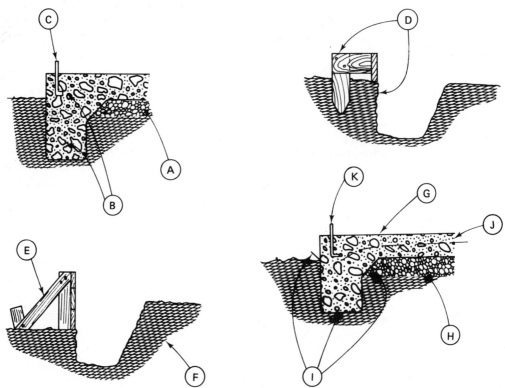

FIG. 6–14 Considerations for a slab with integral footing for a perimeter wall: (a) firm subbase on crushed rock or gravel fill; (b) continuous reinforcement rods; (c) anchor bolts—if needed, footing width should equal 12″ minimum, footing depth should be 18″ minimum or below frost line; (d) excavation in earth *plus* braced boards may be combined as forms; (e) raise boards above grade when necessary; (f) any fill must be solid—well tamped. You should work for ideal (g) slab surface; (h) positive subbase; (i) plastic membrane—continuous; (j) wire mesh reinforcement—plus continuous steel rods; (k) anchor bolts—if needed, for attachment of wood plate.

FIG. 6–15 Be sure to stake all side forms so they are straight, vertical, and rigid. When necessary, additional diagonal bracing can be used to push a board into correct position.

FIG. 6–16 It's a good idea to cut formboard material on assembly. This way you can avoid miscuts because of possible slight errors in mating pieces. Big, visible errors, of course, should be corrected.

FIG. 6–17 Stepping down, in a small stage, can be as simple as this. The lateral piece is the riser board and will be removed as soon as it is practical to do so without damage to the concrete.

FIG. 6–18 After the hole is dug, you can drive down heavy, steel bars or a length of preservative-treated, redwood 2 x 4. The pier hole, plus the reinforcement serve to tie footing and slab together.

FIG. 6–19 This garage floor slab was poured after the foundation work was done. Hence, the only formwork required was at the front where the entrance will be.

edge of the stakes must either be flush with the top edge of the formboards or lower so that projections will not interfere with the screeding work that follows concrete pouring.

There are special considerations for formboards left as a permanent part of the project or, perhaps, to serve as a control joint. These are discussed in Chapter 19.

ANCHOR BOLTS

Anchor bolts are special fastening devices that are set in fresh concrete. They can serve various purposes but they are most commonly used in a foundation pour, where anchor bolts are used to secure the sill to the concrete.

Some craftsmen are rather lax with this phase of the job, and the resultant inaccuracies cause problems later. Do the job right by spacing the anchor bolts correctly and centering them in relation to the sill (attached later) and by setting them vertically. This can be done by eye if you work carefully, but good work will be assured if you set up temporary holders for the bolts. Holders can be made from scrap pieces of 1-in. board, nailed to span the forms and drilled to accommodate the diameter of the bolt. The spacing between them is usually 6 ft, but additional bolts should be used at corners and on each side of door openings. When you hang the bolts in the wooden strips, be sure that the threaded end projects an amount that is equal to the thickness of the sill plus the thickness of the washer and nut plus about ½ in.

Some workers will "float" sills almost immediately after the concrete is

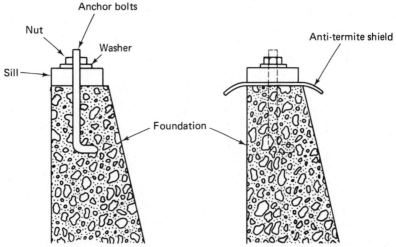

FIG. 6–20 The sill is secured to the foundation with anchor bolts during the pour. The nut and washer must not be tightened until concrete has set enough to take the pressure. The sill—usually made of foundation-grade redwood—may be floated; that is, it may be placed while the concrete is wet. A metal anti-termite shield may be incorporated.

poured. When working this way, cut the sills to correct length and predrill them accurately to fit over the anchor bolts.

FORMS FOR FOUNDATIONS

Forms used for foundations are more difficult than those required for slabs with regard to materials, time, and effort. Foundations (or footings) must be strong and sound to guarantee the life of the structure you put on them. So neglect is not a wise policy.

Walls, and the footings they sit on, may be cast as one piece or they can be cast separately. In either case, the "form" for the base parts is often trenched into the soil itself. The design and depth of the trench must assure that the concrete will be on firm ground and below the frost line. When conditions are average, the width of the footing should equal twice the thickness of the wall; its depth should be at least equal to the wall thickness. These guides can be affected by such conditions as type of soil, structure weight, climate, and the like. In the final analysis, the successful project conforms to all the standards and restrictions that are imposed by local codes.

Don't be negligent of formwork so as to get to the pouring. Pay special attention to the bracing. Although there are minimum requirements, no law

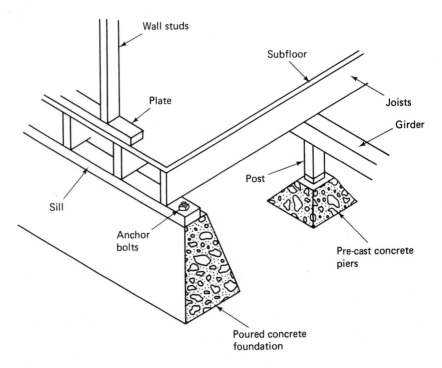

FIG. 6–21 Foundation—crawl space.

FIG. 6–22 Precast piers that include wooden blocks for nailing purposes are available. They should always be set firmly on concrete footings. The piers will take posts or girders.

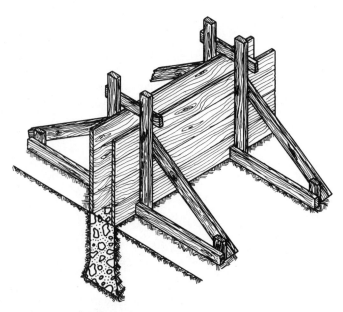

FIG. 6–23 Above-grade foundation walls can be done this way. Earth trench is form for footing. The walls of trench must be straight and true. Since most foundations require a footing, *be sure to check local codes before doing this!*

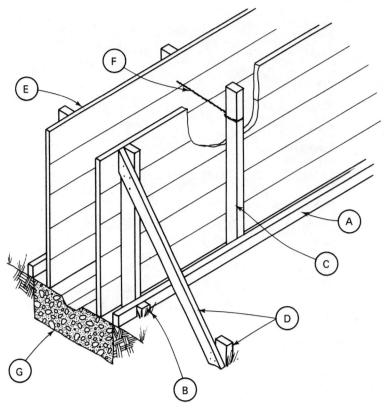

FIG. 6–24 Assembling a form for a concrete wall: 2 x 4 baseboards (a) are secured solidly with 2 x 4 stakes (b); 2 x 4 verticals (c) are spaced about 18″ on centers. Strong diagonal supports and stakes (d) are used. Form boards (e) may be 1 x material or you can work with plywood. Also available are readymade forms that you can rent; these may be of steel or wood. Ties (f) pass through forms and twist around the verticals— use #10 wire. Keyed footing (g) is used.

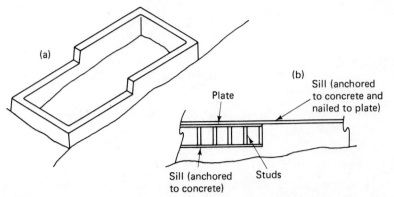

FIG. 6–25 In situations where it is economical to step down a pour (a) the stepped-down section is built up to level by framing (b).

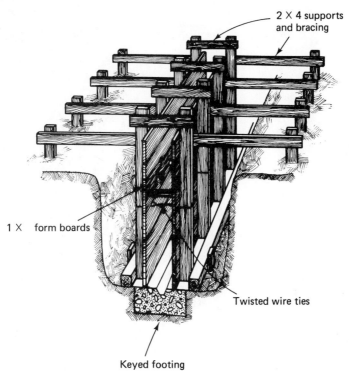

2 × 4 supports
and bracing

1 ×   form boards

Twisted wire ties

Keyed footing

Fig. 6–26 Foundation form in a sub-grade trench.

Fig. 6–27 Stepped down sections (for a doorway?) are simply boxed off in the formwork. This is much better than having to work with a sledgehammer later to achieve the same thing.

FIG. 6–28 Stepped foundations can mean considerable savings in concrete and form materials. This kind of thing is often practical to do when the site is a slope.

FIG. 6–29 The flexibility of concrete makes it possible to cast it in just about any shape. The entire foundation should be viewed as a single piece, even when you are turning corners, as shown here.

FIG. 6–30 It used to be that girder pockets required a good amount of detailed formwork. Today's professionals use blocks of styrofoam and simply knock it out after the concrete is set.

FIG. 6–31 The flexibility of concrete makes it possible to cast it in just about any shape. The entire foundation should be viewed as a single piece, even when you are forming intersections for interior or exterior bearing walls.

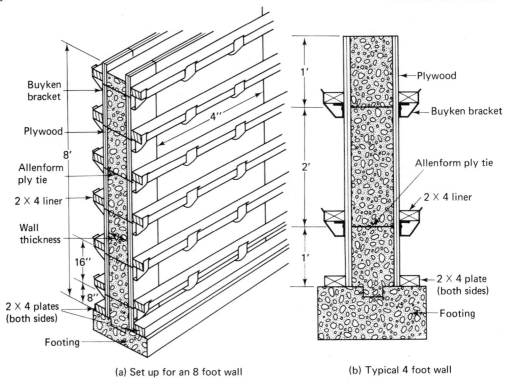

Buyken
bracket

Plywood

8'

Allenform
ply tie

2 × 4 liner

Wall
thickness

16"

8"

2 × 4 plates
(both sides)

Footing

4"

(a) Set up for an 8 foot wall

1'

2'

1'

Plywood

Buyken bracket

Allenform ply tie

2 × 4 liner

2 × 4 plate
(both sides)

Footing

(b) Typical 4 foot wall

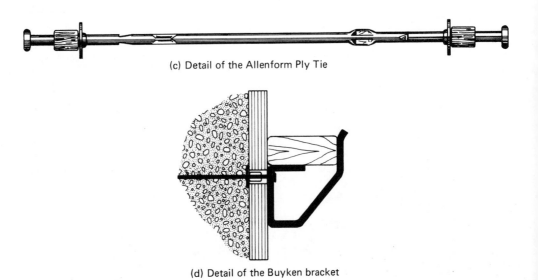

(c) Detail of the Allenform Ply Tie

(d) Detail of the Buyken bracket

Fig. 6–32 The trades utilize reusable materials as concrete forms such as the conform system that combines buyken brackets with allenform ply-ties.

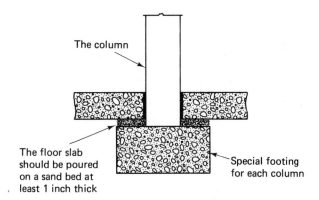

FIG. 6–33 Details for interior columns. Note: Wrap the base of the column with 3 plies of 15 pound felt before the pour.

confines you to any maximum amount. You can, of course, overdo it, but whenever you are in doubt, be generous. A little too much is a lot better than not enough.

The sketches and photographs in this chapter show various designs for good formwork and will acquaint you with a type of commercial material used in formwork. In some areas, this material can be rented. If this is the case where you live, consider that renting can save time and effort worth the rental fee.

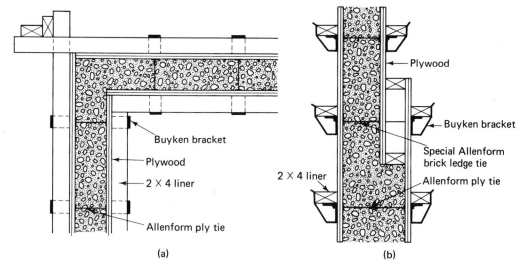

FIG. 6–34 How the buyken bracket with allenform ply-tie system is used to (a) turn a corner and (b) incorporate a brick ledge.

FORMS FOR CURVES

Material that is ½-in. thick is easily bent to form curves for slab work. If the material is not flexible, however, you can employ the "kerfing" method to bend anything from 2 × 4's to plywood sheets. The idea is to make parallel saw cuts across the grain of the wood. Keep the depth of the cuts to about three fourths the thickness of the stock. The number of cuts and how they are spaced depends on the amount of bend required and, to some extent, the material being used. The cuts remove enough of the wood so that the board can be bent back on itself. When the degree of bend varies, make more cuts and space them closer at points of greatest bend. Remember that the cut side is the *outside* of the form.

Other materials such as hardboard, stiff fiberglass, and sheet metal can be used for curve forming. When the material is thin, be sure to support it with adequate bracing.

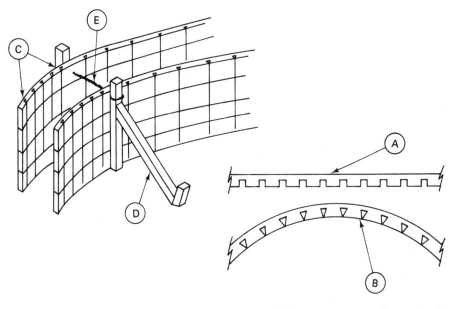

FIG. 6–35 Kerfed boards may be used for curved forms. (a) Kerfs are saw cuts made across the stock. They can be done with a handsaw or on a table saw; (b) kerfed boards bend toward the cut side; (c) sharper bends require more closely-spaced kerfs; (d) use adequate and frequent bracing; (e) wire ties are used.

QUESTIONS

6–1   Generally, how should forms be designed?
6–2   Name some materials that are usable for formwork.
6–3   State some factors that should be considered when choosing formwork material.

FIG. 6–36 Sheet metal may be used for curved formwork. Many times, especially when the turn or curve is extreme, sheet metal is easier to work with than kerfed lumber or plywood.

6–4   What is a *release*?
6–5   Make a drawing that shows the layout of a rectangular building with an attached garage. Show all batter board locations and include placement of lines.
6–6   Describe a method that can be used to cut steel reinforcement bars.
6–7   Which statement is correct?
           When you must join bars for a continuous run, you should:
           (A)   weld the bars together
           (B)   overlap them and tie them with wire
           (C)   butt them end to end
           (D)   buy special, extra-long bars
6–8   Describe wire mesh reinforcement.
6–9   Where should reinforcement materials be placed?
6–10   Make a sketch that shows the amount of overlap and the number of ties required when joining steel bars. Indicate the diameter of the rods and the gauge of the wire.
6–11   How far apart should stakes be placed when you are using 2 in. lumber as forming for a slab?
6–12   What is an *anchor bolt*?
6–13   What is the correct spacing for anchor bolts and where should additional ones be used?
6–14   Make a sketch that shows a cross-section of a concrete foundation wall and footing. Show typical dimensions.
6–15   Name some materials that may be used for forming curves.

7

Dumping and Spreading

Concrete should be placed as soon as possible after it has been correctly mixed. Whether you work with a wheelbarrow or the trough of a ready-mix truck, try to place the concrete as near as possible to its destined location. Dragging or flowing the concrete excessively can overwork the mix and cause water and fine particles to come to the surface. Ignoring this caution can result in dusting and scaling problems later.

Do not overwork the concrete during the dumping and spreading operations. Concrete will fill uniformly and to full depth with minimum attention. Forcing it will, among other things, cause excessive amounts of the coarse aggregate to move to the bottom of the forms.

Spreading can be accomplished efficiently with the square-edged shovel or with a special concrete rake. Often, an ordinary garden rake or hoe is used, but be aware that too much of this can separate large pieces of aggregate from the mortar. Work with the shovel along the forms and in corners so as to eliminate voids. In deep forms (walls) use a length of 2 × 2 or 2 × 4 as a tamper. Some amount of this is necessary to fill the forms, but don't overdo it. Keep tapping the outside of the forms with a hammer. This will help to settle the concrete and will result in a smoother surface.

After the forms are full, the concrete must be leveled to the top edge of the

Fɪɢ. 7–1   Dump freshly mixed concrete as close as possible to its final position. Avoid excessive raking and spreading. Much of the necessary settlement can be accomplished with a square-edge shovel.

Fɪɢ. 7–2   Having an assistant is almost a must when concrete requirements call for delivery by a ready-mix truck. Here, one worker does a preliminary screeding job while the other works with a shovel.

forms. This is the "strikeoff" or "screeding" operation and may be performed with a straight piece of 2 × 4 that is long enough to span the forms. Often, too much concrete is spread before the screeding begins, or a struck-off area is allowed to sit too long before being floated. Both these situations can allow water from the mix to collect on the surface (bleed water). A point strongly emphasized by the profession is that serious scaling and dusting will result from any operation performed on the surface of concrete while bleed water is present. Ignore this fact, and you will be disappointed in the results.

Place the straightedge so that it spans the forms and work it to and fro in short saw-like actions as you move it toward a finish point. Tilt the tool in the direction of travel and, to fill any low spots, keep a small amount of concrete ahead of it. If enough concrete piles up to make the tool difficult to move, just

FIG. 7–3   The mix may be transported by wheelbarrow. Note the bridge that has been erected over the gridwork so the weight of the concrete will not distort the formwork.

FIG. 7–4   This worker is doing the screeding operation using a length of straight 2 x 4. By keeping a trowel nearby, he is able to quickly fill depressions and voids that are visible in the struck area.

remove part of it with the shovel. If low spots appear after you have passed with the straightedge, fill them in and redo the screeding. As with all other concrete-work operations, do only as much screeding as you must to level the pour. If everything else has been done correctly, it is possible for a single screeding pass to do the job.

Screeding is followed by hand-floating, bull-floating, or darbying. The choice of tool depends on the size of the job. Hand-floating only is okay but best confined to very small pours. Generally, one uses a bull float or darby first, followed by a hand float.

These operations level ridges and fill voids that remain after screeding. They also assure that all particles of coarse aggregate will be slightly below the surface of the concrete. The blade of the bull float should be pushed ahead with the front edge raised so as not to dig into the concrete. When you pull the tool back, keep the blade flat. If, at this point, you still have holes or depressions in the surface, shovel on some additional concrete and resmooth with the float.

A good job with a darby requires two passes. Work once with the blade of the tool flat, using a to-and-fro sawing motion. On the second pass, tilt the blade of the tool slightly and move it laterally.

Hand-floating should be done after the concrete has stiffened a bit. This can occur quickly on hot, dry, or windy days or take several hours when the weather is cool or humid. The absence of water sheen and the ability of the concrete to

FIG. 7–5   A jitterbug tamper may be used after the screeding operation. It is used in short, quick, up-and-down motions and assures that particles of coarse aggregate will be correctly settled.

FIG. 7–6   The tamping is followed by bull-floating. This should eliminate major ridges and fill any remaining voids. Tilt the front edge of the blade when pushing forward; keep it almost flat when returning.

FIG. 7–7   Floating and troweling follow the bull-floating. Note that the worker is using walking and kneeling pads. This, so his weight will be more broadly distributed on the fresh concrete.

sustain foot pressure without damage are fairly reliable signs that you can begin the hand-floating.

Hand-floating and troweling take a little practice. The novice creates problems with excessive pressure and by allowing the tool to dig in. Both tools are to be used in a smooth, sweeping arc action with passes that overlap about 50 percent. The wood float leaves a slightly rough texture which may be used as is, especially when good traction and skid resistance are required. Troweling produces a smooth, hard, dense surface and should not be done unless the concrete has already been surfaced with a float.

When you start troweling, make one pass with the blade flat over the entire surface. Apply only as much pressure on the tool as you need to do the job. Be extremely careful to avoid digging into the surface. A single troweling operation may produce a surface you are happy with. If not, go back to the starting point and repeat the operation but, this time, with the blade of the tool slightly tilted.

QUESTIONS

7–1   Describe, in general, the correct procedures to follow when dumping and spreading concrete.

7–2   State, in the correct order, the procedures that follow dumping and spreading.

7–3   How can you judge when the concrete is ready for floating?

7–4   What is the difference between *floating* and *troweling*?

8

Special Finishes

A wood float leaves a slightly rough, gritty surface; a metal trowel produces a harder, smoother finish. These are very common methods of finishing concrete slabs, but the possibilities for creating attractive and unique effects are almost unlimited. For example, you can show striations and swirl patterns, create a flagstone effect, do pseudo or real exposed aggregate, add color, even make the surface resemble wood. It is a good idea to experiment on a small, trial slab. In all cases, remember that the technique must not diminish the strength of the concrete.

BROOMING

The effects that you can get with brooming depend on how soon you apply the broom, whether you use it wet or dry, and whether the bristles are soft or hard. Special brooms made for the purpose are available, but interesting patterns can be achieved with the broom you use on the patio or in the shop.

Broom finishes are attractive and produce nonslip textures that are very useful where sure footing is required. The technique is simple: Pull a damp or dry broom across the concrete while it is still soft enough to be marked. If this

FIG. 8–1   This is the kind of patterned swirl finish you can produce with a hand float. Notice that the arcs are fairly uniform and that the sweeps overlap. The finish looks good and provides good traction.

FIG. 8–2   Broomed finishes can be coarse or quite fine depending on the broom-bristles and the hardness of the concrete when the job is done. Straight pulls with the broom produce this effect.

is done quickly after floating, the texture will be coarse—especially if you work with a damp, stiff-bristled broom. If you follow floating with troweling and then work with a dry soft-bristled broom, the texture will be much finer. Between these extremes various degrees of texture are possible.

Don't push the broom back and forth. Instead, pull the broom toward you, using one-direction, slightly overlapping, parallel strokes. When the technique is used on a sidewalk, or even a driveway, do the brooming so that the texture lines are at right angles to the traffic direction.

Straight lines, sweeping curves, tight waves—all can be accomplished, depending on how you use the broom. You can even create a herringbone or quilt pattern by sweeping adjacent areas so that the texture lines are at right angles.

Be careful with curing procedures that must follow. You don't want to destroy the texture that you have worked to create.

WASHING

The technique of washing is employed to expose some of the mix aggregates, but the resulting surface is not considered to be a bona fide exposed aggregate finish. A broom is used after the floating operation to remove a small amount of the surface materials. This is followed, after the concrete has set sufficiently, by a wash from a garden hose. The spray and the angle at which you direct it affect how much material is removed; it should not be overdone. Excessive brooming and washing will remove too much of the finer materials from around

FIG. 8–3   Brooming is the first step before washing with a hose. The job here is to remove some of the surface material. It can be done rather quickly after floating has been accomplished.

FIG. 8–4   Washing is done with a garden hose. Having someone work with you to broom off material loosened by the water is a good idea. At this point, the concrete is hard enough to take some traffic without damage.

FIG. 8–5   You have some control over the effect you get in brooming and washing by what you put in the concrete mix. You can control the size of the coarse aggregate; also use aggregates of different color.

the pieces of coarse aggregate, so that they can easily be knocked off by normal traffic.

Success depends much on accurate judgment as to when the concrete is ready for the brooming and washing. Do it too soon and the operation can be a disaster—better to be a bit cautious. You can always increase spray force and water volume by adjusting the nozzle on the hose. One good indication that the concrete is ready is when you can walk on it without damage.

To get an effect that more closely resembles true exposed aggregate, you can work with a concrete mix that contains colored aggregate.

EXPOSED AGGREGATE

The most common method of laying the popular exposed aggregate finish is called "seeding" because the surface of a regular concrete mix is embedded with a layer of special aggregate. The concrete is placed, leveled, and floated in normal fashion but with its surface about ⅜ in. below the forms to allow room for the aggregate that will be added.

The special aggregate—and you have a choice of color and size—is then spread over the concrete in a uniform layer. To flatten and start the embedment

FIG. 8–6   After the floating is done, spread a uniform layer of the aggregate you have chosen over the surface of the concrete. Tossing it up and letting it fall will help embed it. Cover all areas.

of the added aggregate, work with a hand float or a darby, using a tamping action. Follow by using a bull or hand float as you would normally until the surface of the job looks like any normal concrete surface after it has been floated.

The next step is to use a stiff-bristled broom lightly to remove any excess mortar. This should not be done until the surface is hard enough to bear the

FIG. 8–7   Do a thorough job of leveling and embedding the aggregate by working with a darby or a hand float. In this operation the tools are used more like tampers. Work over the entire slab.

FIG. 8–8   Follow the tamping operation by working with a hand float until the surface appears as any concrete slab would after being floated. All of the seeded aggregate must be embedded and coated.

FIG. 8–9   Follow floating by working with a stiff-bristle broom. Make light passes to remove excess mortar. Don't start this until the slab is strong enough to support a man's weight without damage.

FIG. 8–10   The final step is to wash and brush until all the aggregate is exposed and free of any cement film. This professional's broom can be fitted to a garden hose so one man can do both watering and brooming.

weight of a man. A good test is to place a small piece of plywood (about 1 ft²) on the surface and then stand on it. If the surface is not indented, it is probably safe to proceed with the next step.

The final operation is to combine fine spraying with brushing. The trade uses a special broom that contains water nozzles, but the chore can be accomplished successfully with a garden hose and a regular broom.

Don't continue if the operation dislodges any of the aggregate. Wait until the concrete is set enough to avoid damage. The washing and brushing operation should continue until the aggregate is uniformly exposed and there is no noticeable cement left on the stones.

TRAVERTINE

Travertine is a very pleasant finish that resembles the texture of travertine marble. To do it, mix a batch of mortar (see Chapter 12) to the consistency of very thick paint and apply it, in what can only be described as a splotchy manner, to the surface of freshly leveled concrete. The idea is to create ridges and depressions. Allow the coating to set a bit and then work with a trowel to spread and flatten the mortar. The texture that results is smooth on raised areas and coarse between them.

The extent to which the surface looks like travertine depends on how much mortar you apply and the amount of troweling you do. It is also possible to point up the effect by adding a coloring agent to the mortar mix.

FLAGSTONE

A flagstone-like finish can be done in either random or regular patterns. To do this you need a special tool that can be made by bending up an 18-in. length of ½-in. copper pipe into a gentle S-shape. A rounded corner of the pipe is used to score the concrete right after it has been floated. The result should be a pattern of concave grooves. Some irregularities will occur along the lines because the tool will move large pieces of aggregate. It is best to hand-float the surface after the first scoring job and then to go over the lines again.

Fig. 8–11 Flagstone effect is a pattern of concave grooves made with the tool described in the text. Actually, the size of the pipe used to make the tool is not critical. It can be anything from ¼-inch to ¾-inch.

Fɪɢ. 8–12 Uniform grooves are often used to break up a large expanse of concrete. These are done with a grooving tool called a cheater and often are combined with deeper control joints.

Fɪɢ. 8–13 Random-spaced grooves plus coarse brooming can make concrete resemble wood. Do the grooves before you do the brooming. Thus, the surface and the grooves will have the same texture.

FIG. 8–14 The tools must be used carefully and at a point when the concrete can take the deep impressions. Unfortunately, as of now, the tools are not available for renting, but a contractor can do the job for you.

STAMPED PATTERNS

You can use something like a coffee can, a pie tin, or a bread pan to make impressions in concrete. Special tools that work in somewhat similar fashion can be used to form patterns that resemble tile, brick, or stone. The special tools cut rather deeply and are applied to partially set concrete. At this writing, such work must be done by a contractor who is franchised to use the tools; they are not available through rental agencies. For more information you should contact a local concrete contractor.

COLORING

Concrete may be colored in one of two ways—by adding a special mineral oxide pigment to the mix itself, or by applying a dry coloring material to the surface of the concrete after it has been floated. In the first method, the pigment is combined with mix materials in the dry state and thoroughly mixed. The amount of pigment should not exceed 10 percent of the weight of the cement. When colored concrete is prepared in batches, all proportions must be carefully controlled so that results will be uniform.

The second method consists of spreading dry coloring material over the

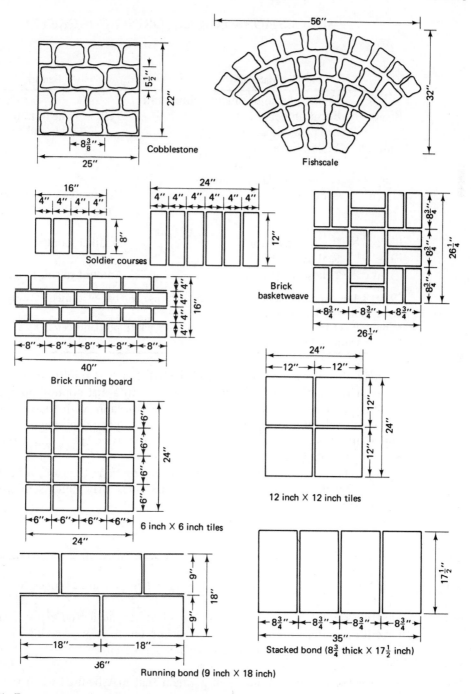

Fig. 8–15   Pattern-stamped concrete provides the advantages of slab construction but with the appearance of brick, or tile, or cobblestones. (The tools that create the patterns shown here are the product of the Bomanite Corp.)

SUGGESTIONS FOR COLORING CONCRETE*

To Get This Color	Use This Material†	Remarks
White	White portland cement, white sand	
Brown	Burnt umber	Yellow oxide of iron
	Brown oxide of iron	will modify the color
Buff	Yellow ocher	
	Yellow oxide of iron	
Gray	Normal portland cement	
Green	Chromium oxide	Yellow oxide of iron will shade the color
Pink	Red oxide of iron	Small amount
Rose	Red oxide of iron	
Cream	Yellow oxide of iron	Amount to suit tone wanted

*9 lb. is the maximum amount of pigment to use per bag of cement. Always read the instructions on the pigment container.
†You will get truer colors when you work with white portland cement and aggregates that are white or light in color.

concrete after it has been leveled and floated. For best results, apply about two thirds of the total amount of coloring material in a first application and do the finishing as you would normally. Repeat the procedure with the remaining coloring material. Uniform spreading and working in of the coloring powder is very important. It is easy to see how you can get a mottled effect unless the powder is equally distributed.

In either case, be sure to read all the instructions that are printed on the coloring agent container. Amounts and methods may differ somewhat from brand to brand.

QUESTIONS

8–1   What are some of the factors to consider when you finish concrete by brooming it?
8–2   Describe the difference between a *washed* finish and a true *exposed aggregate*.
8–3   Describe the steps needed to do a *travertine* finish.
8–4   How can you make a tool to produce a flagstone effect?
8–5   Can concrete be colored? If so, how?
8–6   What is the maximum amount of pigment that may be used when the material is part of the mix?

9

Working with Brick

Bricklaying can be a delightful hobby and a rewarding vocation. The profession recognizes that it is an ancient and honorable craft. Recent excavations have revealed that brick structures existed many thousands of years ago. Brick is durable and attractive and may be used for projects that range from simple walks to intricate load-bearing structures.

The novice cannot rival the professional bricklayer's speed, but he can do a fine job if he works at his own pace. Unlike much concrete work, which involves a do-it-before-it's-too-late factor, bricklaying can be more or less paced to suit the craftsman's time and energy. It is possible, for example, to mix small or large batches of mortar to suit the number of bricks you can, or wish to place, in a given period of time.

Brick has always been an indoor as well as an outdoor material. Originally, it was most evident in kitchens and fireplaces. Now, owing in great part to the hundreds of combinations of sizes, types, and colors that are available, you will find inside brick projects that range from walls and floors to dividers and bars and even built-in furniture. Very special materials include thin slices of real brick or plastic imitators that can be applied to existing walls with an adhesive or mastic. Often, such materials make it possible to remodel without concern about weight.

FIG. 9–1   Brick is just naturally compatible with the outdoors and will harmonize with just about any kind of landscaping. This brick has been laid in a "running bond."

FIG. 9–2   Innovative design can cause dramatic effects. This brick was laid in a conventional running bond pattern but with slight projections that create dark shadows. Appearance changes with movement of the sun.

FIG. 9–3   The monotony of this high, long wall is broken by incorporating tiles in the design. This kind of thing can also be done to improve air circulation in, for example, closed courts.

Brick can be very formal, rustic, or dramatic. It may be used strictly for utility purposes or for decorative touches. It all depends on the type of brick, its size, its color, the method of assembly, and the design of the joints.

BRICKLAYING TERMS

ALL STRETCHER BOND—bricks laid end to end but with all the vertical joints staggered

BED—the joint between the courses of brick; also refers to the base that supports the structure

BED JOINT—on a horizontal plane, the joint between courses

BOND—the arrangement that provides strength to the structure plus an attractive pattern

BREAKING THE JOINTS—bricks laid so that the vertical joints in courses are not in line

BRICK VENEER—a wall, one brick thick, that covers another structure

BULLNOSE BRICK—designed with rounded ends or corners

BUTTERING—the act of placing mortar on the brick before setting it in place

CLOSER—a brick, broken to a size that fits a small space

CLOSURE—a brick that is broken in half lengthwise; in effect, half a header brick

COMMON BOND—a pattern of stretcher courses but with a header every fifth or sixth course

COPING—the top course of a masonry wall

COURSE—a single, horizontal row of bricks

FACE—the long, narrow side of a brick

FULL HEADER—brick laid across a wall

GROUT—a rich, but thin mortar that will flow easily into cracks and joints for filling (used to fill cavity walls)

HEADER—same as "full header"

LAP—the length of one brick that rests on the next one

POINTING—the material used to fill a joint in brickwork; the act of filling a joint

ROWLOCK—a course of headers that are laid on edge

SET—a special chisel used for cutting brick (brick set)

TUCK-POINTING—pointing done on existing masonry

WALL TIES—special metal reinforcements used to provide strength where two walls meet (also for cavity walls)

QUESTIONS

9–1   What does *bond* mean?

10

Types and Sizes of Brick

To avoid confusion, it is best to place available brick types into one of four basic categories that are generally accepted by building supply dealers. All brick sold by reputable dealers meets the standards that have been established for the trade.

Face brick is probably the highest-quality product. Rigid manufacturing standards assure that texture and color will be uniform and that each unit will be nearly perfect. It will be difficult to find such defects as flaking, chipping, cracking, and warpage.

Common brick (often called *building brick*) is as strong as face brick but standards permit more imperfections. For one thing, the texture and the color will not be as carefully controlled.

Fire brick is also a very high quality product that is made from special clays. This brick is excellent for fireplaces because it holds up regardless of tremendous heat.

Paving brick, sized for use without mortar joints, is highly resistant to cracking under great loads. The brick gets a longer baking period and, like fire brick, it is made from special clays.

WEATHERING

Brick is durable but can be damaged by freezing weather. The degree of resistance that a brick has under such conditions relates to the amount of water trapped inside it. Longer baking periods cause more clay fusion and this reduces the water absorption capacity of the brick. Control of the baking time makes it possible to produce several types of brick that are good for various weather situations:

SW (severe weathering) has the highest resistance to freeze–thaw, rain–freeze conditions.

MW (moderate weathering) can take some rain–freeze conditions but not severe ones.

NW (no weathering) is good for use in mild climates where no freeze or even hard frost conditions are possible.

Common brick is available in any one of the three grades. Facers are available as SW or MW; pavers and fire bricks are SW. The decision as to what type to select can be made for you by local building codes. When you have a choice within the code restrictions, choose on the basis of the appearance you want and the cost of the brick. A factor to consider is whether the brickwork is interior or exterior. Protected, inside areas will often permit the use of less expensive units. The same is true of brick that is used as fill material and for backup walls.

USED BRICK

Used brick is quite popular, but few people realize that the choice may not be sound structurally and that the units, authentic or imitation, are more costly than new brick. Salvaged brick may have pores so permeated with impossible-to-remove mortar that a fresh mortar joint will be 50 percent as strong as it should be. Old buildings contain both high-quality and low-quality units and it is not likely that salvagers will sort out the two. Brick that was manufactured 50 years ago cannot compare with the quality of units made today.

In conclusion, salvaged bricks are a risk whenever the project must bear loads but are acceptable for decorative purposes and for brick veneering. *Pseudo* used brick is actually new brick that has been deliberately abused, usually in a tumbler of some sort, and then discolored. It is easy to spot since the patina that only time can supply is absent, but the overall effect is acceptable. At any rate, you can be sure that new used brick is structurally sound.

FIG. 10–1   Manufactured "used" brick is structurally sound and has the appearance of salvaged units. They cost more than regular brick because of procedures that must follow normal production.

MODULAR CONSIDERATIONS

Bricks that are designed for modular assembly work out in multiples of 4-in. These dimensions make provision for joints that occur on ends, surfaces, and backs. The width of two bricks matches the length of one, and joint thickness is considered so that the overall assembly will work out in uniform fashion. Nonmodular bricks don't work out the same way; it is often necessary to use a partial brick to attain uniformity with a mating course or to conform with a joint pattern.

When you are ordering supplies from a dealer, be sure that you are talking the same language. The *nominal* size includes allowances for mortar joints; the *actual* size does not.

SCR BRICK

SCR (Structural Clay Research) means a particular kind of brick with nominal dimensions that are 2⅔ × 6 × 12 in., but the overall significance of the letters applies to a special masonry process that was devised for the trade. The system provides for increased efficiency, better workmanship, and more production in

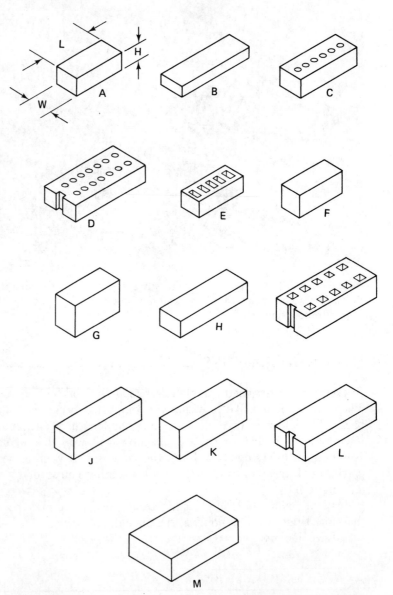

FIG. 10–2 Common brick sizes (modular). Cores shown in sketches are merely representative. Some units are available either solid or hollow. Permitted tolerances may account for varying sizes in actual products.

	Name	Nominal size			Joint thickness	Actual size		
		W	H	L		W	H	L
A	Standard	4	$2\frac{2}{3}$	8	$\frac{3}{8}$	$3\frac{5}{8}$	$2\frac{1}{4}$	$7\frac{5}{8}$
					$\frac{1}{2}$	$3\frac{1}{2}$	$2\frac{1}{4}$	$7\frac{1}{2}$
B	Roman	4	2	12	$\frac{3}{8}$	$3\frac{5}{8}$	$1\frac{5}{8}$	$11\frac{5}{8}$
					$\frac{1}{2}$	$3\frac{1}{2}$	$1\frac{1}{2}$	$11\frac{1}{2}$
C	Norman	4	$2\frac{2}{3}$	12	$\frac{3}{8}$	$3\frac{5}{8}$	$2\frac{1}{4}$	$11\frac{5}{8}$
					$\frac{1}{2}$	$3\frac{1}{2}$	$2\frac{1}{4}$	$11\frac{1}{2}$
D	"SCR"	6	$2\frac{2}{3}$	12	$\frac{3}{8}$	$5\frac{5}{8}$	$2\frac{1}{4}$	$11\frac{5}{8}$
					$\frac{1}{2}$	$5\frac{1}{2}$	$2\frac{1}{4}$	$11\frac{1}{2}$
E	Engineer	4	$3\frac{1}{5}$	8	$\frac{3}{8}$	$3\frac{5}{8}$	$2\frac{13}{16}$	$7\frac{5}{8}$
					$\frac{1}{2}$	$3\frac{1}{2}$	$2\frac{11}{16}$	$7\frac{1}{2}$
F	Economy 8	4	4	8	$\frac{3}{8}$	$3\frac{5}{8}$	$3\frac{5}{8}$	$7\frac{5}{8}$
					$\frac{1}{2}$	$3\frac{1}{2}$	$3\frac{1}{2}$	$7\frac{1}{2}$
G	Double	4	$5\frac{1}{3}$	8	$\frac{3}{8}$	$3\frac{5}{8}$	$4\frac{15}{16}$	$7\frac{5}{8}$
					$\frac{1}{2}$	$3\frac{1}{2}$	$4\frac{13}{16}$	$7\frac{1}{2}$
H	Norwegian	4	$3\frac{1}{5}$	12	$\frac{3}{8}$	$3\frac{5}{8}$	$2\frac{13}{16}$	$11\frac{5}{8}$
					$\frac{1}{2}$	$3\frac{1}{2}$	$2\frac{11}{16}$	$11\frac{1}{2}$
I	Jumbo (6)	6	4	12	$\frac{3}{8}$	$5\frac{5}{8}$	$3\frac{5}{8}$	$11\frac{5}{8}$
					$\frac{1}{2}$	$5\frac{1}{2}$	$3\frac{1}{2}$	$11\frac{1}{2}$
J	Economy 12	4	4	12	$\frac{3}{8}$	$3\frac{5}{8}$	$3\frac{5}{8}$	$11\frac{5}{8}$
					$\frac{1}{2}$	$3\frac{1}{2}$	$3\frac{1}{2}$	$11\frac{1}{2}$
K	Triple	4	$5\frac{1}{3}$	12	$\frac{3}{8}$	$3\frac{5}{8}$	$4\frac{15}{16}$	$11\frac{5}{8}$
					$\frac{1}{2}$	$3\frac{1}{2}$	$4\frac{13}{16}$	$11\frac{1}{2}$
L	Norwegian (6")	6	$3\frac{1}{5}$	12	$\frac{3}{8}$	$5\frac{5}{8}$	$2\frac{13}{16}$	$11\frac{5}{8}$
					$\frac{1}{2}$	$5\frac{1}{2}$	$2\frac{11}{16}$	$11\frac{1}{2}$
M	Jumbo (8)	8	4	12	$\frac{3}{8}$	$7\frac{5}{8}$	$3\frac{5}{8}$	$11\frac{5}{8}$
					$\frac{1}{2}$	$7\frac{1}{2}$	$3\frac{1}{2}$	$11\frac{1}{2}$

general masonry construction. Quite often, it is possible for a single row of SCR bricks (correctly used) to be used in place of a two- or three-brick-thick wall of conventional bricks. Such applications must conform to local codes.

COLOR AND TEXTURE

Color and texture are matters of personal taste. A good procedure is to visit a local supply yard and actually see and feel what is available. Many suppliers will have pseudowalls set up so that buyers can judge the appearance of various materials in construction state. Short of this, there is nothing wrong with stacking a dozen bricks in bond fashion and then making a decision. The joints won't be there but you will gain some idea of final appearance.

QUESTIONS

10–1   List the names of the four basic types of brick.
10–2   List the letter designations of brick and tell what they are.
10–3   Give two reasons why it may not be wise to use salvaged brick for structural purposes.
10–4   What is the difference between *nominal* and *actual* size?
10–5   What does SCR stand for?

11

Tools for Brickwork

A number of special tools are made for the brick craftsman. Some are essential no matter the degree of involvement; others are nice to have; still others may not be that useful to you. The chances are that you have applicable tools on hand if you are now involved in shopwork, house framing, or outdoor building jobs. Carpenter's squares and levels, conventional flexible tapes, folding rules, lines, line levels, plumb bobs—all will serve your purposes. In fact, many a small brick-laying job has been accomplished with little more than a shovel, a trowel, and a length of line. This is not said to downgrade an assortment of good equipment but only to point out that your needs will be relative.

TROWEL

The mason relates to a trowel the way a cabinetmaker does to a hammer. Manufacturing quality, size, weight, flexibility of the blade are all important. But given a choice of samples with similar characteristics, the worker selects the one that has the right "feel" to it. The blade must hold up under rough usage and considerable abrasive action, so cheap, bargain-counter tools are out.

Trowels are available in different sizes and shapes. Those that have long, narrow blades are more difficult to handle than those with shorter, wider blades,

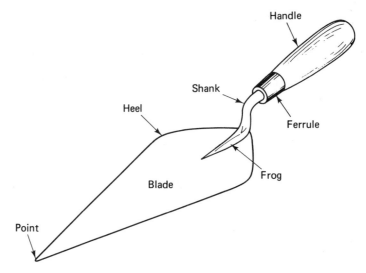

FIG. 11–1 Nomenclature of a trowel.

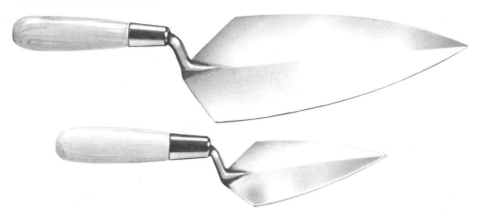

FIG. 11–2 The smaller pointing trowel is used mostly for redoing damaged mortar joints, a process often referred to as "tuck-pointing." Among other special trowels are duck bill, buttering, and cross joint.

because the longer ones hold a lot of mortar weight farther from the wrist. This can be tiring, especially for the novice. A good rule-of-thumb is to buy the largest trowel that you can handle.

BRICK HAMMER

The mason uses the brick hammer for many things. One end is shaped like a chisel and is used by the experienced worker with a fine touch to smooth and shape cut bricks. The opposite end of the hammer is usually square and is used, among other things, for breaking bricks, striking brick sets, and driving nails.

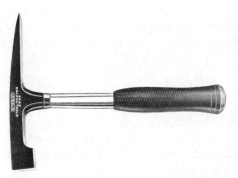

Fɪɢ. 11–3 Bricklayer's hammers come in different styles, but all of them resemble this design. The square end is for breaking bricks and some conventional hammer applications.

There are many head styles and even more options in the shape and materials of the handle, the latter running from wood to nylon. You'll choose one that feels good in your hand, but stay in the 24-oz weight category.

Take good care of the hammer and always be sure that the head is fitted tightly to the handle. Brick cutting and shaping can throw off sharp chips of material, so safety goggles are not an undue precaution.

CHISEL

A chisel may be called a *blocking chisel*, a *mason's chisel*, or a *brick set*. It is used to make sharp cuts on brick or to score bricks that will be broken with the hammer. The handle and blade are made of a single piece of steel and run 7 to 8 in. in overall length; the blade width should be 3 or 4 in. The blade is beveled to achieve the cutting edge required. When necessary, the edge should be dressed on a grinder to keep it in good cutting order.

Fɪɢ. 11–4 Brick sets are wide chisels with a beveled cutting edge. In use, the chisel is held with the beveled edge facing the mason. Maintain the edge if you wish the tool to do a good job.

The striking end of the chisel must also be maintained to keep it free of burrs. Continued striking on a surface so deformed can cause sharp pieces of metal to fly off. Maintain the tool; wear safety goggles when you use it.

RULE

Most masons have a folding rule that extends to 6 ft and is marked on one side to call out brick course heights. Different marks are used in relation to the size of the bricks and the thickness of the mortar joints. The spacing rule is used

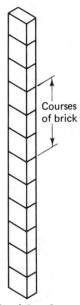

Courses
of brick

FIG. 11–5 A story pole or gauge is a length of 1 x 2 marked off to indicate the height of the brick courses.

FIG. 11–6 Many masons prefer wooden levels with metal reinforced edges to all metal designs. One reason given is that wood feels better in the hand regardless of the weather.

FIG. 11–7 A pocket level is handy for making checks in confined areas or even on individual bricks. This design has a clip so it may be carried about much like a pen.

Fig. 11–8   This is a combination plumb bob and chalk line reel. It contains 100 feet of line and a refilable chalk chamber. The line is coated as it is pulled from the case. The small handle is used to retract the line.

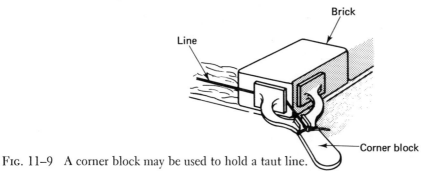

Fig. 11–9   A corner block may be used to hold a taut line.

Fig. 11–10 Brick handlers or "tongs" are used by the trade to facilitate brick handling on the job. They lock and release automatically and are adjustable to carry from 6 to 10 bricks.

directly on the job and for marking off a special *story pole* that is made for particular dimensional callouts.

QUESTIONS

11–1   List four essential tools that a bricklayer should own.

12

Construction with Brick

Strong masonry that is durable and will resist rain penetration is the result of good design, good materials, and dedicated workmanship. Design, in this area, refers more to how the individual bricks are assembled for a particular job than it does to appearance. Appearance, of course, is affected by workmanship, since even in the most common of assemblies, careless attention to uniformity of joints, levelness of courses, and the like, makes for obvious errors.

The design of a masonry project includes the interlocking relationship between bricks and between courses of brick. This "structural bond," plus full mortar joints, should result in the total assembly of many small units as a single, strong structure. Any brick assembly, and this must include the mortar joints, produces a pattern. Flexibility in design permits decorative effects that do not diminish structural strength. Such effects may be achieved by the bond itself or by introducing areas of different color or texture. The effect may be a pattern achieved with brick placement that does not conform to the basic bond, or it can be a series of small, separate, repetitive details. Anyone who desires a preview of an idea can make a small, scaled drawing of the project and then shade the bricks that will form the pattern. The drawing also makes a wise work plan since the interjection of the pattern bricks must occur at precise intervals.

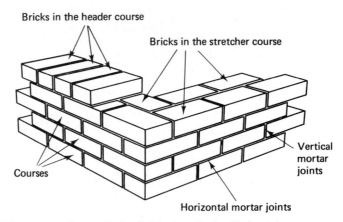

FIG. 12–1 Typical brick construction—all the bricks interlock; all the joints are staggered.

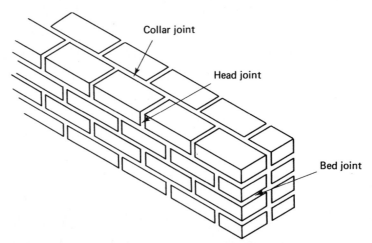

FIG. 12–2 Joints in brickwork.

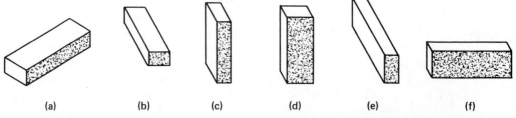

(a)   (b)   (c)   (d)   (e)   (f)

FIG. 12–3 Special names given to bricks because of position in the project: (a) stretcher—end to end with narrow edge forward; (b) header—used as a tie, end is visible; (c) soldier—stands on end with narrow edge forward; (d) sailor—used like the soldier but with broad side forward; (e) rowlock—used like a header but resting on narrow edge; (f) shiner (or rowlock stretcher)—used like a stretcher but resting on narrow edge.

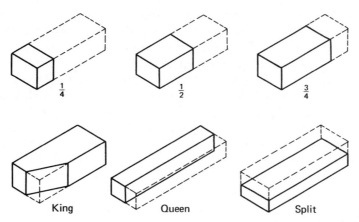

Fig. 12–4 Technical names for bricks that are cut for use as closures and for corners. (Solid lines equal piece to be used.)

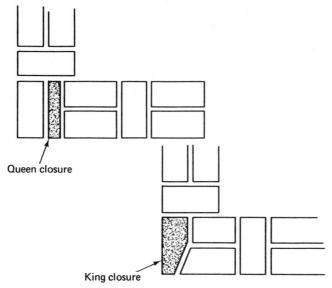

Fig. 12–5 How king or queen is used to turn a corner with English bond. Note that queen is never used as the corner.

MOISTURE IN BRICK

Brick has a considerable amount of capillary action and this can suck enough water from the mortar to cause very weak joints. The absorption rate varies considerably for different types of brick and is judged in the trade by the amount that enters 1 in.^{2} of a brick's surface in one minute when the brick is immersed in $\frac{1}{8}$ in. of water.

To avoid the sucking action that will weaken the mortar joint, brick must be thoroughly wetted and then allowed to surface-dry before being used. A mason's on-the-job test for determining the moisture content of brick can be conducted by anyone. Mark a circle about 1 in. in diameter on the surface of the brick that will contact the mortar. Place 20 drops of water in the circle and note the absorption rate. If it is less than one and one-half minutes, the bricks require wetting.

Bricks may be wetted by immersion but a better way is to wet an entire pile at once by using a garden hose. Spray the pile until water runs from all sides. Water on the surface of the brick can be as bad as no water inside the brick, so allow units to surface-dry before you place them.

WORKING IN COLD WEATHER

The *National Building Code* states: "All masonry shall be protected against freezing for at least 48 hours after being placed. Unless adequate precautions against freezing are taken, no masonry shall be built when the temperature is below 32° Fahrenheit on a rising temperature or below 40° on a falling temperature, at the point where the work is in progress. No frozen materials shall be built upon."

Utter structural failure or, at least, leaky masonry walls result when the mortar freezes before it has set, or when materials and partial constructions have not been adequately protected during cold weather. There is no getting around this if you wish to do a good job and a safe job. Either schedule your work for kind weather or go through the business of heating the materials and keeping the job protected for the minimum amount of time. To supply continuous work for masons and to achieve on-schedule completions, the trade will enclose the project with plastic sheets that contain heat introduced through, for example, gas or oil heaters. Consult local codes in your area for specifics relative to working in cold weather.

RAIN PROTECTION

Partial constructions should be protected from becoming saturated by rain. Otherwise, the job may take many, many weeks to dry out and efflorescence may appear on surfaces during the drying period. Efflorescence is the appearance of soluble salts that are present in mortar and masonry units. Usually it appears as a white powder but it can be a combination of greens and browns. It may be cleaned off, but it is much better to avoid or, at least, to minimize its appearance.

A good way of doing this is to cover the job each time you leave it with a waterproof membrane. Use a sheet that is large enough to cover your materials completely or that will overhang at least 2 ft on each side of the wall. If the weather is breezy, or threatens to be, weight the sheet down to keep it in place.

BONDS

The word "bond" may refer to *structural bond, mortar bond,* or *pattern bond.* Basically, structural bond describes the interlocking systems used when assembling brick projects, but it also covers the use of metal ties and grout in certain applications.

Most structural bonds are based on variations of *English bond* or *Flemish bond,* both of which are tried-and-true methods. English bond has alternating courses of stretchers and headers, whereas Flemish bond includes headers and stretchers in *each* course. The design of the latter is such that headers and stretchers in alternate courses appear in vertical lines.

In any method, the stretcher bricks provide strength longitudinally, whereas the header bricks bond the project transversely. As a general rule, enough headers should be included in the design to comprise not less than 4 percent of the total surface. The distance between headers, vertically or horizontally, should not exceed 2 ft. There may be some variations of this rule even in some building codes but it is important to appreciate the strength provided by headers in any project that is more than one brick thick. Designs that do not include headers, such as *running bond* and *stack bond,* are usually restricted to specially reinforced projects or to use as veneering.

Variations of the basic bond or the incorporation of a detail from another bond are often done to create a pattern or to add a degree of distinction to a

FIG. 12–6   The running bond has continuous stretchers in all courses. The design is used mostly for walls that have special reinforcement and for veneering.

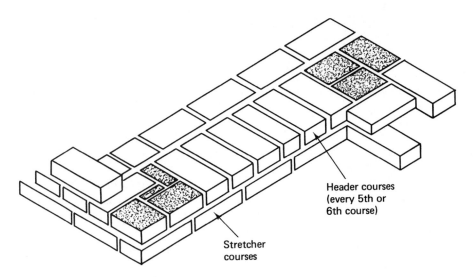

Header courses
(every 5th or
6th course)

Stretcher
courses

FIG. 12–7 How to do a 12-inch thick wall (common bond). Shaded units are closures.

project. *Common bond* (often called *American bond*) includes header courses
at regular intervals, but the pattern can be varied by using a Flemish header
course. To be different with Flemish bond, you can use more stretchers between
the headers in each course. Sometimes a variation becomes a standard and acquires its own name. When three stretchers alternate with a header in Flemish
bond, the design is called a *garden-wall bond*. If only two stretchers are used, it
is called a *double-stretcher garden-wall bond*.

Other variations of basic themes can be achieved by introducing bricks of
different texture or color, by incorporating recessed or projecting units, even by
leaving openings through the omission of units. In all cases, moving from the
norm must not decrease needed structural strength. When in doubt, consult
first with the local building inspector.

FIG. 12–8 How to turn a corner when doing the English bond. The shaded bricks are closures.

BRICK JOINTS

Joints are either finished with a trowel or they are tooled with a special instrument that compresses and shapes the mortar. In the troweled joint, excess mortar is struck (cut off) and the joint is finished with the trowel itself. For other joints, you can work with commercial tools or some that you can make yourself. Joints can be designed in line with a particular effect you want, but one thing they all have in common is that they should be watertight, especially on outside projects.

	Name	Remarks
	Concave	Formed with jointing tool — resists rain penetration — very good in heavy rain and high wind areas — joints are usually kept small
	"V"	
	Weathered	Regarded as a top grade troweled joint — sheds water easily
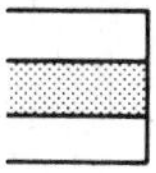	Flush (rough cut)	Easiest of the troweled joints to do — watertightness can't be guaranteed
	Struck	Not difficult to do with a trowel — a common joint but the ledge does not shed water easily
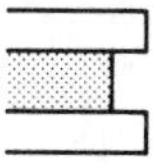	Raked	Formed with a tool — creates good shadow lines — is not particularly resistant to freezing, high winds or rain
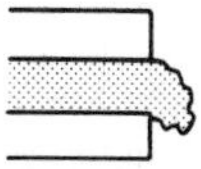	Extruded	Mortar squeezed out is simply left as is — used for appearance only — makes for a very rough wall
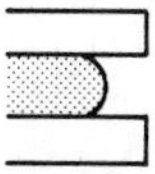	Shaped	A "raked" joint done with a special tool — characteristics similar to the raked joint
	Convex	Extruded mortar is shaped with a special tool — makes a fairly acceptable joint but is done mostly for appearance

FIG. 12–9 Types of mortar joints.

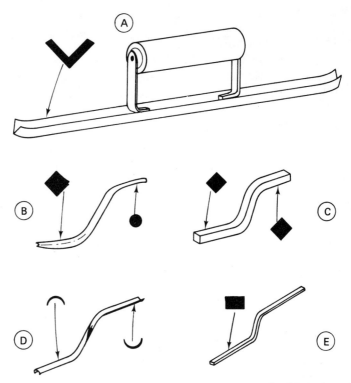

FIG. 12–10 Types of commercial joint tools: (a) sled runners come in 14 inch and 20 inch lengths for either a V-joint (shown) or a half round; (b) combination V and round; (c) V; (d) concave—for raised half round joints; (e) double-blade caulker—often called slicker.

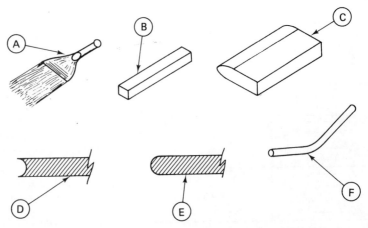

FIG. 12–11 Typical homemade tools for finishing mortar joints: (a) whisk broom—good for swept and raised joints; (b) hardwood or metal bar—for raked joint—even for V-joint; (c) hardwood—shaped for V-joint; (d and e) hardwood—for shaping joint after raking; (f) bent metal rod—can indent after the joint is finished flush. Use a trowel to do the weathered, flush, or struck joints.

FIG. 12–12 You can do some improvising when finishing mortar joints. This worker is using a short length of wood to get a raked effect. Be sure, however, that the work is done uniformly.

Joints will be easy to do if you have proceeded correctly to that point. There will be ample mortar of correct consistency for tooling. To test if the mortar is ready for tooling, see if it will take a thumbprint. Remember that the mortar must be compressed as well as shaped.

MORTAR

Mortar has many uses. Structurally, it must bond all individual units together, and it must completely fill all the spaces between units. It must bond to any reinforcement materials so that they will be an integral part of the structure. Mortar also provides compensation for dimensional variations in units and, of course, it is part of the overall decorative effect.

Mortar must be plastic if you are going to work with it easily and get solidly filled joints. Basically, that is why the lime is added to the mix. If you are interested enough, mix two small batches of mortar, one with lime and the other without. The difference in consistency and handling will be obvious. Cement, of course, is the ingredient that supplies the strength and the set-quick factor.

Sand is included in the mortar to provide body and volume. Since it is inert, it is not affected by the water in the mix, but very fine sand does result in a weaker mortar because it increases water requirements for acceptable plasticity. Good mortar sand is well graded for the purpose and consists of strong, durable, mineral particles, minimal amounts of alkaline and saline materials, and no foreign objects such as large stones and dirt.

Use water that you would not hesitate to drink and as much of it as you need to bring the mix to a plastic, workable state. Blend all the materials thoroughly while they are in a dry state. Since the cement, sand, and lime are of

BRICK MORTAR MIXES*

	Parts by Volume				
Type	Portland Cement	Hydrated Lime	Sand (maximum)†	Strength	Use
M	1	¼	3¾	High	General use where high strength is required, especially good compressive strength; work that is below grade and in contact with earth
S	1	½	4½	High	Okay for general use, especially good where high lateral strength is desired
N	1	1	6	Medium	General use when masonry is exposed above grade; best to use when high compressive and lateral strengths are not required
O	1	2	9	Low	Do not use when masonry is exposed to severe weathering; acceptable for non-load-bearing walls of solid units and interior non-load-bearing partitions of hollow units

*The water used should be of the quality of drinking water. Use as much water as is needed to bring the mix to a suitably plastic and workable state.

†The sand should be damp and loose. A general rule for sand content is that it should not be less than 2¼ or more than 3 times the sum of the cement and lime volumes.

different colors, you'll know that mixing is adequate when the batch is uniform in tone. Add water gradually and continue to mix as you do so. A good mix will hold together but will spread easily. You should get a smooth finish when you stroke across its surface with a trowel. The weight of a brick on a mortar bed plus some gentle tapping with the handle of the trowel should be sufficient to fill joints and squeeze out excess mortar. Right off, there should be enough suction created so that it requires a real sharp tug to remove a freshly placed brick.

Use the mortar as soon as possible after it has been mixed and don't mix more than you can use in under two hours. It is not a good idea to add water (retemper it) to a batch of mortar that has been sitting for more than two hours. Actually, if you mix batches that you can use up in a reasonable time, retempering won't be necessary. If you must do it, use the minimum amount of water that will bring the mortar back to a workable condition.

A good batch of mortar can be mixed in a wheelbarrow and used directly from it. Or you can transfer a few shovelfuls at a time to a mortarboard (you can use a piece of plywood about 1½ to 2 ft²). To keep the basic batch (in the wheelbarrow) fresh and plastic, mix occasionally.

FIG. 12–13 Mortar can be mixed directly on a board when small amounts are needed. Doing it in a wheelbarrow gives a larger source of supply. Don't mix more than you can use in about two hours.

Be careful when working from bulk materials that you proportion ingredients correctly. Don't be too generous with the water. When you are doing small jobs, you can work with ready-mix materials available in sacks. All you have to do is add the water.

QUESTIONS

12–1   Tell three factors that result in strong masonry construction.

12–2   How can the moisture content of brick be determined right on the job?

12–3   What are *stretcher* and *header* bricks?

12–4   What is the general rule to follow concerning the number of headers?

12–5   How should headers be spaced?

12–6   What easy test can you apply to judge if mortar is ready for tooling?

12–7   What materials are used in a brick mortar mix?

12–8   How much water should be used in the mix?

12–9   What are the correct proportions of materials required for a mortar that has high strength?

12–10   What is the general rule for sand content in a mix?

12–11   Does the quality of water required for a mortar mix differ from the requirements regarding a concrete mix?

13

How to Lay Brick

Bricklaying is a precise craft. There are some areas, such as spacing, leveling, and alignment, where you get help from levels, lines, and plumb bobs, but the actual placing of the individual units is a talent you must acquire with practice. The experienced craftsman wastes no effort and no motions; all materials are on hand and well prepared. On the other hand, this is an activity where you can do a good job as you learn. The difference between the novice and the professional has to do primarily with speed.

Begin by imitating the pro: Grip the trowel as he does. Hold it by the blade in your left hand with the handle up and toward you. Grasp the handle with your right hand so that your thumb is topside and on or near the ferrule. Keep your first finger opposite the thumb on the underside of the handle and curl the other fingers around so as to stabilize the tool in your hand. Grip firmly but not as if you were trying to squeeze the air out of a rubber ball. Flexible wrist action is essential, so stay relaxed enough to manipulate the trowel freely. First-time users will feel a strain in the wrist and hand rather quickly, but don't allow this to cause a change in how the trowel is gripped.

One way to get the feel of the trowel is to work the mortar mix with it. Pick up a trowelful and drop it in a straight line on the board. Work the mortar from the edges to the center of the batch; cut into it with the edge of the blade. Always keep the same grip.

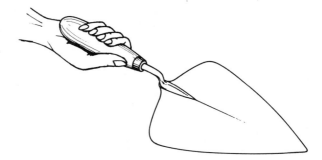

FIG. 13–1 The right way to hold a trowel for "throwing" mortar is with the palm up.

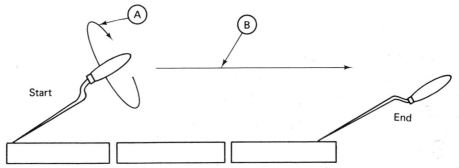

FIG. 13–2 "Throwing" a three-brick mortar bed. Laden trowel is held (with palm up) as shown at the start of the stroke. It is turned (a) through the stroke (b) until the end. Total turn of the trowel equals 180°.

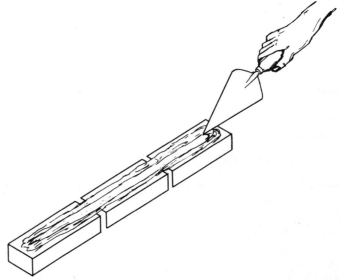

FIG. 13–3 At the end of the stroke, the mortar is spread evenly along the center of the bricks.

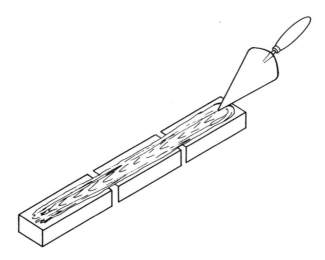

FIG. 13–4   The point of the trowel is then used to spread the center throw as evenly as possible over the bricks.

You can practice laying a mortar bed by placing four or five bricks, stretcher fashion, on a flat surface. Pick up a quantity of mortar and in a smooth motion try to cover more than one or two of the bricks. This throwing action involves turning the trowel as you move it longitudinally. It should deposit enough mortar on each brick so that the furrowing procedure that follows (see sketch) will produce a fairly uniform mortar bed that covers all surfaces. Cut off excess mortar and place it, if needed, on the new mortar bed or return it to the batch. Laying a mortar bed by spot-placing very small amounts of mortar and then spreading them is not recommended.

The head joint (vertical) is done by placing mortar on the end of the brick

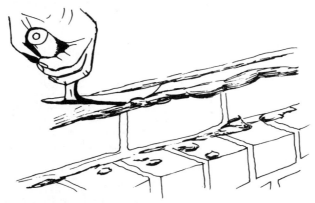

FIG. 13–5   The trowel may be used in this fashion to furrow the bed joint. DO NOT furrow too deeply since it can result in leaky walls.

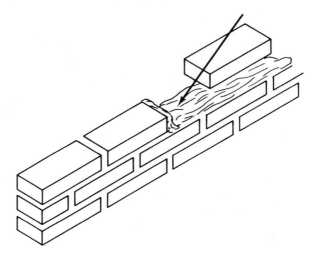

FIG. 13–6 When adding a new brick, apply pressure *down* and *toward* the mating brick in the course.

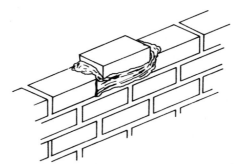

FIG. 13–7 The last brick in a course is often called a *closer* or *closure*. Butter it generously.

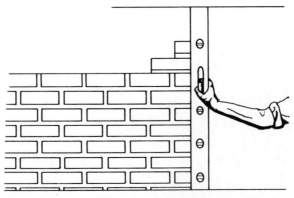

FIG. 13–8 Use the level frequently, especially at corners, to be sure that the project is plumb. Off-line construction is bad visually and structurally.

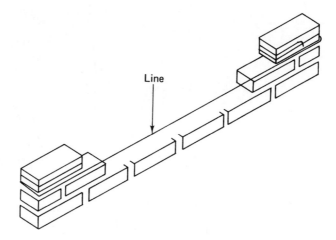

Fig. 13–9 Stretch line taut from built-up corners to establish level of each course of bricks.

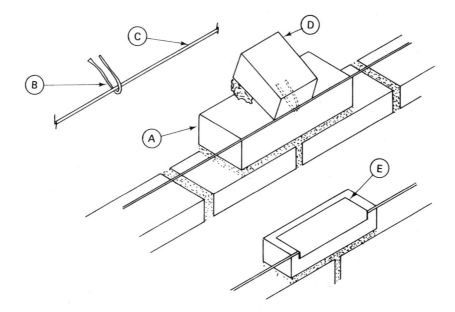

Fig. 13–10 Maintain alignment and prevent sag in a long line. Establish a mid-point brick, called a *trig brick* (a), in correct position. Then loop a small piece of line, called a *trig* (b), around the main line (c) and hold it in place with a piece of brick, tilted and secured with a blob of mortar (d). You can make a trig plate by bending a right-angle flange on a piece of sheet metal (e). Weight the plate down with one or two bricks.

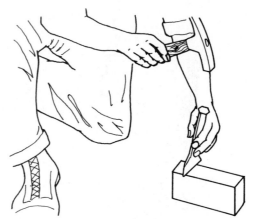

FIG. 13–11 The best way to cut brick is to score it on four sides with the hammer and chisel and *light* taps. A sharp rap with the hammer, preferably on a broad side, should finish the job.

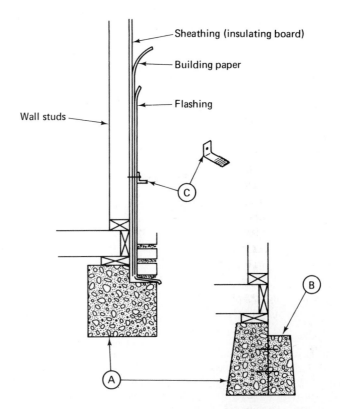

Sheathing (insulating board)

Building paper

Flashing

Wall studs

FIG. 13–12 Brick veneering is supported by a special ledge that is part of the foundation (a) or is added to it by remodeling (b). Special metal ties are nailed to the wall (c) so that the corrugated flange sits in the mortar joint every few courses.

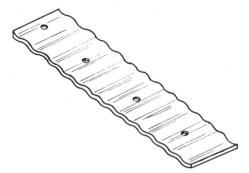

FIG. 13–13 Corrugated metal tie for brick veneer—22 gauge galvanized steel, ⅞ inch wide. Generally, one tie is required for every two square feet of wall. They must be spaced, horizontally and vertically, not more than two feet apart. Use 8d (minimum) galvanized nails and be sure that the nails drive into studs.

before you set the brick on the new mortar bed. The action is down and toward the adjacent brick and should result in firm contact with the bed, and in a full head joint. Tapping with the handle end of the trowel is an aid in setting the brick and in achieving alignment. This action should be minimal and done quickly. If you must rely on heavy hitting to get a brick in line, you have placed it very badly to begin with.

The expert works with a flourish that is hard to follow by eye. You will do so as well, if you wish, but don't be too ambitious at the start. Although your target may be a bed joint to cover four or five bricks, be happy with two or three to begin with. At any rate, don't let a desire for speed cause you to overlook the fact that all joints between bricks must be *full*.

It doesn't hurt to go through a dry run before you actually apply mortar. This means placing the first course of bricks on the footing or foundation with correct joint spacing. If you wish, you can use small pieces of plywood of appropriate thickness as gauges between the bricks. This will point out whether full bricks will fill the line or whether you will have to cut. At this point, when design permits, you can make adjustments in the number of bricks or compensate slightly with joint thickness if it eliminates having to cut small closures. When the dry run proves out, use a pencil or crayon to mark the brick locations on the foundation. Then you'll know exactly where to place them when you start mortaring. A good beginning here is a big step toward successful work.

Start laying bricks at corners, working up as many as four or five courses before you start filling in between. Control course heights at the corner by working with a rule or a story pole; check plumbness with a level. With the corners built up, you have extreme points between which you can stretch line to give you alignment and height for the in-between bricks.

Work carefully to keep joints uniform. A single, small error is tolerable but if you multiply it by 12 or 24 or whatever number of bricks are involved, it can add up to an unhappy event when the moment arrives to fit the last few bricks.

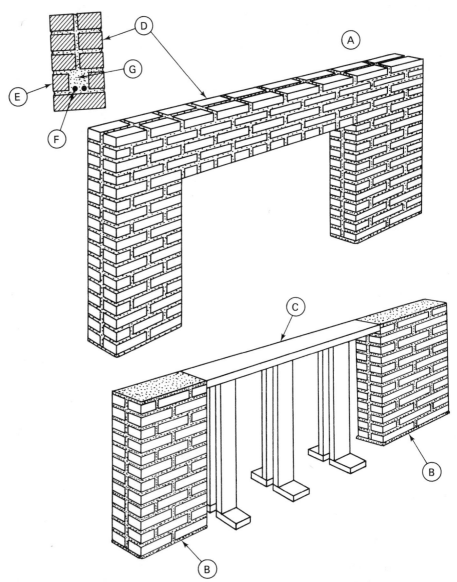

Fig. 13–14 Basics of a reinforced brick masonry lintel, used for an opening or over a door or window. The complete structure (a) is built by first building up the piers (b) and then using 2 x material to assemble a support structure (c). The design may differ but it must be strong enough to support the top brickwork and be easy to knock down. Note that the total height of the support includes mortar bed for the next course of bricks. (d) is a section through the top structure. Bricks are split (e) to provide cavity for two lengths of ⅜ inch reinforcement rod (f) and mortar grout (g). NOTES: Provide about 6 inch bearing over the piers for the reinforcement rods and be sure the project has enough strength to stand alone before removing the support structures—24 hours is an absolute minimum.

QUESTIONS

13–1   Demonstrate how a trowel should be held.
13–2   Demonstrate how mortar should be picked up from the mortar board and spread on the brick.
13–3   What type of mortar placement is *not* recommended?
13–4   Why is it wise to go through a *dry* run before actually placing mortar?
13–5   What is a *story pole*?
13–6   Why is joint uniformity very important?
13–7   Make a sketch that shows how a line is used as a guide for a course of brick.

14

Pseudobrick

It is perhaps unfair to use the word "pseudo" to refer to materials that look like brick but are not, because some of the materials we have in mind are 100 percent masonry and will withstand extremes in temperature inside and outside the house. Others are not so usable and must be restricted to particular areas. In general, they differ from conventional brick in the way that they are installed and because they are not designed for use as structural, loadbearing walls.

Without exception, pseudobrick comes in veneers, made for attachment to existing subbases whether they are vertical (walls) or horizontal (floors). How these veneers are attached depends on the particular product. Some require a very adhesive mastic, others a mortarlike substance. Often, assuming that the product is the proper one for the use you have in mind, you can base your choice on the installation procedures. For example, the units of one product are simply applied to an overall coverage of mastic. The mastic that appears between the bricks is the joint. With others, spaces between bricks are filled with a special compound after the bricks are set.

The bricks may be separate pieces or they may be mounted on panels composed of several courses, several feet long. The panels interlock and attach to studs; a few nails can put up a lot of brickwork.

We show here but a few examples of the many products that are available. Installation procedures vary and may be very critical in relation to a particular product. Always read and follow the instructions that come with the materials you buy. Be sure that your choice is right for your purposes.

FIG. 14–1   Such units may be attached to fairly lightweight subbases. In this case, ⅜-inch plywood over 2 x 4 framing was used. The first step is to apply a full coating of special mastic.

FIG. 14–2   Then, the back of each piece of brick is buttered generously with the same mastic. The buttered brick is set into position with a very slight, sliding action.

FIG. 14–3 Specially shaped pieces are provided so you can turn inside or outside corners. The spaces between the pieces are not filled so the appearance is that of a raked joint.

FIG. 14–4 This product is made by the Eldorado Stone Corporation and is 100% masonry. It's made to withstand extreme temperature inside or outside the house. It's difficult to tell from the real thing.

FIG. 14–5 Apply a generous amount of mortar to the back of the brick. Be sure that all areas are covered. You can test the consistency of the mix by applying a brick and checking the holding strength.

FIG. 14–6 The Eldorado product may be applied directly to clean, untreated surfaces, but it will probably be necessary to place metal lath over paint. On wood surfaces, an application of roofing paper is recommended.

FIG. 14–7 Apply the brick firmly and with a gentle jiggling action to insure a good bond. Don't slide the bricks into position or you will remove too much mortar.

FIG. 14–8 Use the applicator as you would a cake decorator. Adjust pressure so that the bead of mortar you squeeze out is just enough to fill the joints. This part of the job may be done with regular bricklaying tools.

FIG. 14–9 The same company makes stone-like materials that are applied in similar fashion and have the same durability as the brick. Check local supply sources for actual views of what is available.

FIG. 14–10 Let the mortar dry a bit and then use the striking tool (also part of the kit) to clean and smooth out the mortar bead. You can finish by brushing the joints with a whisk broom.

15

Concrete Block

Concrete blocks are cast masonry units that can be made with many different types of aggregate. Sand, crushed stone, and gravel are familiar ones. But such materials as volcanic cinders, expanded slag, and specially treated shale or clay may also be used. In a literal sense, concrete block refers to units that are made with the crushed stone or sand and gravel aggregates, but the name has come to include all hollow masonry units made with any of the materials mentioned.

The choice of aggregate contributes to whether the block is "light" or "heavy." A block measuring 8 × 8 × 16 in. can weigh as much as 50 lb when cast with gravel, crushed stone, or slag. Other materials that are naturally light in weight (volcanic cinders, scoria, pumice) can produce blocks that weigh between 25 and 35 lb. Since either type can be used for many kinds of masonry construction, a choice can be based on texture, availability of types or shapes in a particular area, local codes, or insulation factors. Generally, lightweight units do a better insulating job. Often, the cores of hollow units, whether they are lightweight or heavyweight, are filled with a granular insulating material.

Concrete block, in general, has greater sound-absorbtion powers than, say, a smooth, dense wall of concrete, plaster, or glass. In this area, too, the lightweights are more efficient.

Not too long ago, concrete masonry units were used solely for prosaic proj-

FIG. 15–1 Pierced units (screen block) are available in different sizes and designs. Most times they are used in a stack bond since this repeats the unit design without interruption.

FIG. 15–2 It's difficult to call this a concrete block but such materials are part of the world of masonry units. This is a type of slump block that looks much like adobe. Note affinity with heavy beams.

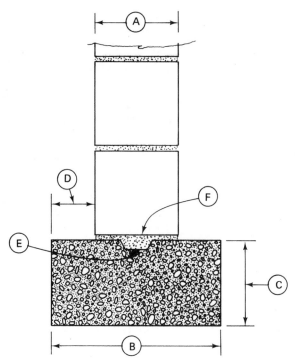

FIG. 15–3   Suggested dimensions for footings (in relation to the wall): (a) represents the thickness of the wall; (b) equals 2a; (c) equals a; (d) equals ½a; (e) is the "key" formed with beveled 2 x 4; (f) is the bed of mortar—full.

FIG. 15–4   Reinforcement procedures are determined by the type of block and by the design of the project. Steel rods of various sizes, like those described for concrete work, are used.

Fig. 15–5 Horizontal steel may be called for in some situations and with particular types of block. How much steel and its placement are factors subject to local codes.

ects and were not considered to make any visual contribution. Today, because the adaptability of the basic product is recognized and because more manufacturers are producing unique and attractive units, concrete block rivals the growth of all other modern building materials.

Like concrete and brickwork, masonry unit construction is subject to building codes that are set up to assure durable projects with built-in safety. Such regulations should never be ignored, but equally important is the fact that codes can provide you with professionally produced construction details that are exactly right for the area in which you live. This includes sizes of footings, the degree of necessary reinforcement, and so on.

TYPES AND SIZES

Like brick, unit sizes of block are usually listed in nominal dimensions. The 8- × 8- × 16-in. unit actually measures less in each dimension by ⅜ in., which is the thickness of the mortar joint. Other sizes are called out in similar fashion: A half unit is 8 × 8 × 8 in.; a half-high stretcher is 4 × 8 × 16 in. Thus, all can be incorporated in a structural or design pattern with uniform results.

Both the heavyweight and the lightweight units are available in the shapes that we show in the sketch, but be aware that other sizes and shapes may be obtained. Practically every manufacturer of block has exclusive shapes and, possibly, block made of different material. Such block does conform to the usual dimensions but appearance may vary a good deal in the effort to produce something that has special charm.

To learn more about what is available, schedule a trip to the storage yard

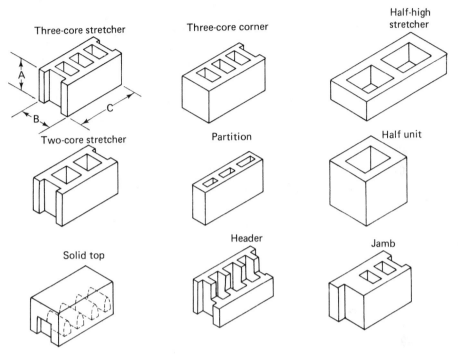

FIG. 15–6   Shapes of common masonry units. Modular unit size (a x b x c) = 8 inches x 8 inches x 16 inches. Thus, a half unit would equal 8 inches x 8 inches x 8 inches; a half-high stretcher would equal 4 inches x 8 inches x 16 inches; and the partition block would have the same dimensions as the half-high stretcher.

of one or more suppliers. It isn't necessary to be limited to what is on display. The supplier cannot possibly stock everything that is available, but he should have a collection of manufacturers' catalogs that tell a more complete story.

The world of masonry units covers more than what we have come to accept as concrete block. Units may resemble oversized bricks that are solid or cored. Some are partially faceted so that a particular pattern results when a certain number of units are placed in a group. *Split block* has a ready-made, rustic facing. *Slump blocks* result in a rugged-looking project.

Pierced blocks (*screen block*) are increasing in popularity. Indoors or out, they serve in practical capacities as dividers or screens and the like. As a patio screen, for example, they provide a good deal of privacy with minimum blockage of air movement. Even large projects appear light and airy, without the security-wall look that so many of us associate with concrete block material. Assemblies of screen block are seldom used as major, load-bearing components in a structure, even though the design is clever enough to make them appear so.

Whatever the job you plan, be sure to check out special units and shapes that may be incorporated to facilitate a given procedure or to end a project in a professional manner. Such items would include *header blocks* for insetting floor

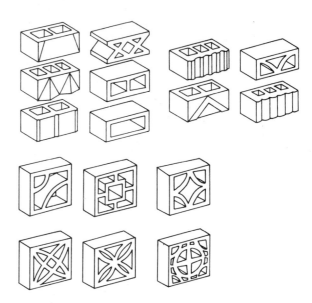

Fɪɢ. 15–7 Decorative block units. In addition to pierced units, there are those with recessed or raised areas that are used to create sculptured wall patterns.

joists; *cap* units that will finish off the top of a wall, a foundation, or even a planter; and corner blocks for ending a wall.

PATTERN BONDS

Bonds are just as important in concrete block as they are in brickwork. Many effects are possible and choices may be based on structural design, local climatic conditions, architectural considerations, and individual preferences. It is easy to see that if you have an assortment of modular units you can create various designs by combining the units in different ways. Colors, texture, units with surface sculpturing, all may be considered when planning visual effects.

The cores in units are often utilized as design features. Uniform spacing of blocks placed at a 90-degree angle can produce attractive openings in a wall. This can be done sparsely or with alternating blocks; units may be set on the long or the short dimension. Stack bonds, where all joints are aligned vertically and horizontally, are as usable in block masonry as they are in brickwork but, here too, such assemblies usually call for special reinforcement procedures.

You can plan the effect you get by doodling on paper in the manner we described for brick. Do remember that the *pattern bond* (for looks) does affect the *structural bond* (for strength), so that a superappearance doesn't always go along with adequate strength. The purpose of the project has, of course, some bearing in this area, but to be safe, have your idea checked out in terms of compliance with local codes. Judgments will always be made in relation to structural

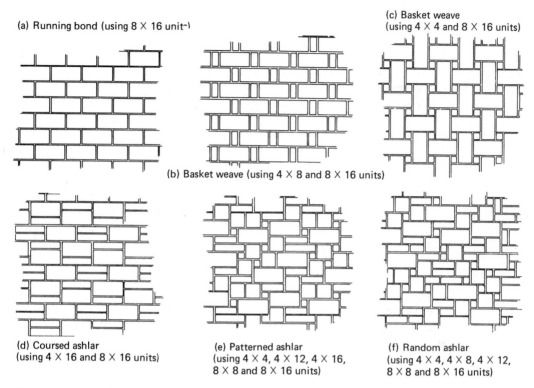

(a) Running bond (using 8 × 16 units)

(b) Basket weave (using 4 × 8 and 8 × 16 units)

(c) Basket weave
(using 4 × 4 and 8 × 16 units)

(d) Coursed ashlar
(using 4 × 16 and 8 × 16 units)

(e) Patterned ashlar
(using 4 × 4, 4 × 12, 4 × 16,
8 × 8 and 8 × 16 units)

(f) Random ashlar
(using 4 × 4, 4 × 8, 4 × 12,
8 × 8 and 8 × 16 units)

FIG. 15–8   Just a few of the patterns you can create when doing concrete masonry.

FIG. 15–9   Three-dimensional effects are created by using flat-surfaced units in this fashion. This can be done with half or whole units. Strong, changing shadow lines produce a dramatic design.

considerations. It is very possible that the design you envision may be unwise as is but permissible with the addition of reinforcement materials.

JOINTS

Joints should be weathertight and neat, with sharp clean lines. Unless there are special reasons for other designs (having to do with appearance), the joints should be tooled in either concave or V-shaped fashion. Don't wait too long before working the joints. A good test, as with brick, is to allow the mortar to set just long enough so that it will take a thumbprint. At this point, the tooling operation will compact the mortar and force it tightly against the masonry on either side of the joint.

You can form good V joints with a tool made from a length of ½-in.-square bar stock and good concave joints with a tool made from ⅝-in.-diameter bar stock or heavy tubing. In each case, the leading edge of the tool should be bent up in a gentle turn. This prevents gouging the mortar when the tool is moved along a joint line. Some masons prefer a tool that has a bend at each end because it enables them to work the tool in two directions. Because common masonry units are rather large, it is a good idea to make or to buy jointing tools that are at least 22 in. long. Vertical joints are formed with a small S-shaped tool.

Joints can be accentuated with deep tooling (raked) if this blends with the design of the pattern bond. It is even possible to emphasize the horizontal joints by raking them and doing the vertical ones flush. The point is to blend the vertical joints with the units as much as possible. To do this, cut off most of the excess mortar after it has hardened a bit and rub the remainder flush with a wad of burlap. The technique can be used to produce strong horizontal lines on every course or on every second, third, or fourth course, depending on the effect you want. A more massive appearance results when there is more space between lines.

The *extruded joint* is strictly a matter of appearance and is not guaranteed

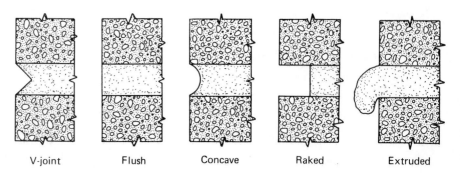

| V-joint | Flush | Concave | Raked | Extruded |

FIG. 15–10 The most common mortar joints for concrete masonry.

Fig. 15–11 A long tool, bent up at least at the leading edge, is used to form joints. The tool is made from square or round rod depending on whether you want a concave or a V-joint.

Fig. 15–12 The vertical joints are done in similar fashion but with a smaller tool. Do the tooling as soon as the mortar has set enough to take a thumbprint. Horizontal joints are done first.

to be waterproof. It is formed by excess mortar that is squeezed out when the units are placed. The mortar is simply left to harden.

Tooling of joints will leave small burrs of mortar. These can be trimmed off with the edge of a trowel or wiped off with a piece of burlap.

MORTAR

Good mortar, correctly applied, will bond masonry units into a strong, durable, well-knit project. All the considerations mentioned for the correct mixing of mortar for brick apply to the mortar you make for block. Severe frost conditions and severe stresses call for a more durable and stronger mortar than is required

SUGGESTED MORTAR MIXES FOR CONCRETE BLOCK WORK*

Type of Work	Cement	Hydrated Lime	Mortar Sand (measured in damp, loose condition)
Routine projects	1 part of masonry cement	None	2–3
	or		
	1 part of portland cement	1–1¼	4–6
Projects subject to extremely heavy loads, severe frost action, earthquakes, violent winds; also for isolated piers	1 part of masonry cement and 1 part of portland cement	None	4–6
	or		
	1 part of portland cement	0–¼	2–3

*All proportions measured by volume.

under mild conditions and ordinary service. Since both these factors are difficult to describe for all areas, it is wise, as always, to check local codes to see if the mortar-mix suggestions made in the chart are correct for your area. Ask particularly about any precautions you should take when working in cold weather.

Mortar that has stiffened because of water evaporation can be brought back to a workable, usable state by thorough remixing and by adding as little additional water as you need to do the job. Mortar can also stiffen because of hydra-

Fig. 15–13 Experienced masons will often pick up a trowel full of mortar and then give it a sharp snap. This removes excess that might drop off and be lost. It also forces the mortar to stick to the trowel.

Fig. 15–14 Stockpiling of units should be done on a platform that prevents contact with the ground. When necessary, place a cover over the stack to keep the blocks from getting wet.

tion (the setting action). When this happens, the material should be discarded. To avoid these problems, work by the clock. Don't mix more mortar than you can use in under two and one-half hours if the air temperature is 80° F or higher. If the temperature is below 80°, you can add about an hour to the time limit.

CONDITION OF BLOCKS

Although masonry units are very hard, careless handling can chip off corners or cause complete breaks. Thin webs on some screen block and details of sculptured units are particularly vulnerable. All blocks should be carefully transported to the site and maintained in their original good condition by careful stacking, preferably on heavy planks raised above the ground.

Block should be delivered in a dry state and used that way. Unlike brick, the material is not wetted before use. Dry blocks combine directly with a good mortar mix for optimum bonding strength. You must therefore protect against wetting by covering stockpiles with a tarpaulin or something similar, as well as the tops of projects between work periods. This will help prevent rain or snow from entering the cores. The degree of protection will depend on weather conditions in your area.

You must also protect blocks from becoming dirty and dusty. Such films will prevent good adhesion with the mortar.

QUESTIONS

15–1   Name some of the materials that are used in cast masonry units.
15–2   Tell the nominal dimensions of a typical block.

15–3   Are the nominal dimensions referred to above the actual dimensions?

15–4   What is a *pierced* block (screen block)?

15–5   Name the two joints that are recommended for block work.

15–6   Name two materials that you can use to make tools that will form the above joints.

15–7   List the ingredients for *two* mortar mixes that can be used for heavy-duty block work.

15–8   Should block be wetted before use?

15–9   How should you protect blocks?

16

How to Lay Block

The tools and most of the procedures that we described for bricklaying apply to blocklaying also. Since the units are rather heavy, it is a good idea to spot stockpiles so that lifting and carrying will be minimized. Grip the units firmly and aim for nearly perfect placement. Excessive jiggling and moving of the blocks to achieve alignment does not help you get a good mortar joint.

You'll notice that the face shell is thicker on one side. This extra thickness should always face *up*, as it will provide a greater amount of bedding area for the mortar. A full mortar bed covers all web areas; a partial one (*face-shell mortaring*) does not. Partial mortaring will suffice for most work. For example, use a full mortar bed for the first course of blocks regardless of whether you are erecting on a footing, a foundation wall, or a concrete slab. Use face-shell mortaring on the remaining courses.

A dry run—placing the first course of blocks corner to corner without mortar —is very important. Check for correct spacing and true alignment. It is much easier to make corrections *now*. It is also possible, design permitting, that some adjustment might eliminate the need to use a cut block to fill out a line.

With the free blocks in place, you can trace around them with a crayon or use a chalk line to mark the footing. Also mark the position of each block to facilitate accurate placement when you start to work with mortar.

FIG. 16–1 Supports for concrete block work can be footings, cast foundation walls, or concrete slabs. The base should be done in line with good craftsmanship and must conform with local building codes.

FIG. 16–2 Many contractors who do a considerable amount of concrete block wall in a single operation will use temporary bracing as an added safety factor. It's wise when back-filling must be done quickly.

Fig. 16–3 Steps to waterproof the exterior surface of the wall are easy to do when the units are so exposed. The system is called "parging." The procedures are explained at the end of this chapter.

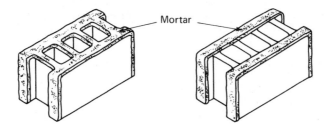

Mortar

Fig. 16–4 Full mortar bedding is shown at the left. All vertical and horizontal surfaces are coated. This is always done with first-course blocks. Most times, face-shell mortaring (right) is used for the following courses.

FIRST COURSE

Start the first course by placing a full bed of mortar on the footing so that you can set down two or three corner blocks. Don't attempt to confine the mortar to the dimensions of the block; instead, be a little generous so that furrowing with a trowel will supply plenty of mortar for contact with all the bottom edges of the face shells.

Place the first block carefully with enough pressure to embed it firmly; check it with a level on both vertical and horizontal planes. Following blocks should be buttered on the ends only. You can accomplish this by setting several blocks on end and applying mortar to the top edges. This is the shell that will contact the adjacent block.

Pick up a block and position it with a motion that applies some pressure both down on the mortar bed and against the mating unit. Do this with at least

FIG. 16–5 Do a dry run as a first step. Align the blocks carefully and be sure that the spaces between the blocks are uniform. Small pieces of plywood, ⅜-inch thick can be used as temporary spacers.

three blocks and then immediately work with a level and by gentle tapping with the handle end of the trowel to make sure that the blocks are correctly aligned. Do the same thing at an opposite corner and then fill in the first course.

One caution that applies generally is the following: Don't apply bed mortar or butter block ends too far in advance of actual placement. Use the trowel to cut off excess mortar that is squeezed from the joints and return it to the mortar supply. A few strokes of the trowel will work it back into the mix. Remember

FIG. 16–6 Start the job by applying a generous coating of mortar to the base. Spread and furrow to be sure that all surfaces of the block will make good contact with the mortar.

FIG. 16–7 Place blocks in the bed mortar carefully and firmly enough to squeeze out excess mortar. Action of placement should be down and toward the block that is already in place.

FIG. 16–8 The vertical joint-edges of the block can be buttered as shown here. How many you prepare at one time depends on how fast you work. The beginner should start with one or two.

FIG. 16–9 It's very important to insure that the blocks in the first course are plumb. Carefully placed blocks will require little adjustment. Don't break the mortar bond.

FIG. 16–10 Work this way to check the levelness of the first course. The level you use should be long enough to span three blocks. Use the level longitudinally and laterally.

that good joints are accomplished when the mortar is soft and plastic. Making block adjustments after the mortar has stiffened is taboo.

CORNERS

Work at each corner to build up courses that will then be used as guides to establish height and alignment for the blocks that go between the corners. It is obvious that perfect work should be your goal. Build up corners four or five courses high, checking constantly on all planes to be sure that the blocks are in alignment and that the structure is plumb.

Bed mortar for the second, third, and following courses is usually applied only to the horizontal face shells of the block. The vertical face shells are done

FIG. 16–11 Let's point out here that bed mortar is applied to the top, outer edges on each course of block. The block to be added is buttered on the vertical face shells only.

FIG. 16-12 Build up the corners to a height of at least 4 or 5 courses. The extent of height and length of corners depends, of course, on the overall size of the job. Check frequently with the level on surfaces.

as they were in the first course. You can, if you wish, do the buttering of the vertical shells on the block that has already been laid. Some masons do both, feeling that this is the way to guarantee full joints.

You can work with a story pole when you are building up the corners. Like the one used for brickwork, it is just a straight piece of wood (1 × 2 will do) marked off to indicate joints and course heights. When marked off correctly, it does much to eliminate the possibility of human error that exists when course heights are established by frequent checking with a flex tape or folding rule.

FIG. 16-13 Check vertically on adjacent corners. Use the level as a straight edge when adding block.

FIG. 16–14 Special L-shaped corner blocks are made for walls that are thicker than 8-inches. It's possible that such a shape may not be available in your area. If so, you can work as shown in the next two photos.

FIG. 16–15 Use a conventional 8-inch x 8-inch x 16-inch block as the corner. Fill the inside corner with a piece of solid concrete brick.

FIG. 16–16 Set the fill-in piece in place after the corner is organized. Butter the three contact edges generously and slide it in place as shown here.

BETWEEN CORNERS———————————————————

Stretch a line from corner to corner for each fill-in course of block. The line will indicate exactly where the top, outside edge of each block must be. Placing the block correctly is important. You should establish good contact with the bed and face mortars as you achieve the alignment indicated by the line. Through practice you will develop a procedure that is comfortable for you. Just be sure to handle the block so that you don't hide the lines of the course below or have to push the guide line aside. Most masons will tip the block a bit and make initial contact with the rear edge of the structure. Then a slight rotating action

FIG. 16–17 With the corners organized, you can set up a mason's line to establish height and alignment of the following courses. Stretch the line taut. Set it up as described for brickwork.

FIG. 16–18 Handle the blocks so they may be set without disturbing the line. Many masons will make initial contact at the rear edge and with a slight sliding action toward the block that is already in place.

FIG. 16–19 In some areas, full mortar bedding may be required on all courses. This means applying mortar to cross-webs as well as edges. Check local codes.

combined with a gentle shove against the adjacent block will place the block so that only a minimum adjustment to the mason's line will be required.

Although we have talked about partial mortar applications on blocks in courses that follow the very first course, be aware that this method is not acceptable in all areas. Local codes may specify that full mortar treatment be afforded each block. This means that on each unit, you must place mortar on crossweb surfaces as well as on perimeter face shells.

Use extra care when placing the closure block. Apply mortar generously to all four vertical edges. Lower the block carefully into position to avoid knocking off the mortar. If you have any reason to believe that you have lost mortar or if there is evidence of open spaces in the joint lines, remove the closure block and redo the job.

FIG. 16–20 Butter all contact edges of the blocks that are already in place in addition to the normal buttering of the block to be added. Too much mortar, here, is better than too little.

INTERSECTING WALLS_____

An interlocking masonry bond is *not* used for an intersecting *bearing* wall unless it occurs at a corner. The end of one wall butts against the face of the other with a control joint used at the point. Special, ¼-in. thick pieces of metal (tiebars) are included in the joint to tie the walls together, thus providing lateral support. The tiebars are shaped so that they have 2-in. right-angle bends at each

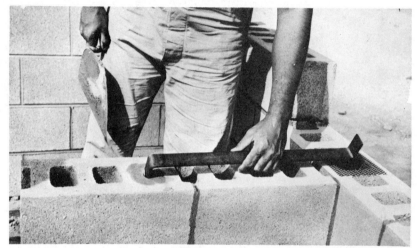

FIG. 16–21 Special steel bars are included in the joint between intersecting bearing walls. No masonry bond is used unless the joint occurs at a corner.

FIG. 16–22 Note how metal lath is used to supply a "floor" for the cores that will be filled with concrete or mortar. Space the tiebars not more than 4-feet vertically.

Fig. 16–23 Reinforcement material in the joint between nonbearing walls can be metal lath or ¼-inch galvanized hardware cloth. When the walls are built together, the reinforcement is placed as shown here.

end. These ends are embedded in block cores filled with concrete. Spacing between the tiebars (vertically) should not extend beyond 4 ft. The dimensions of a typical tiebar in this case equal ¼ × 1¼ × 28 in.

Nonbearing block walls are tied in with strips of metal lath or ¼-in. galvanized hardware cloth. Pieces of the cloth are cut to correct size and placed across the joint between the two walls in every other course. If the intersecting wall is to be constructed later, the metal lath or mesh is included in the appropriate courses of the first wall.

Fig. 16–24 When the intersecting nonbearing wall is to be added later, the reinforcement material is included in the joints of the first wall. Use it in every other course.

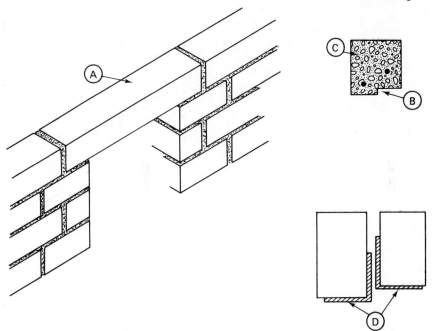

FIG. 16–25 Openings over doors and windows are usually spanned with a precast lintel (a). When the opening is for modular units, the precast lintel is made with an offset (b). (C) is the exterior face. Lintels can also be made by supporting block on steel lintel angles (d).

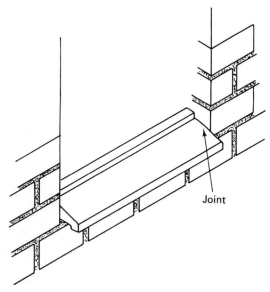

FIG. 16–26 Common procedure is to place precast sills after the walls are up. Pack joints tightly with mortar or a caulking compound.

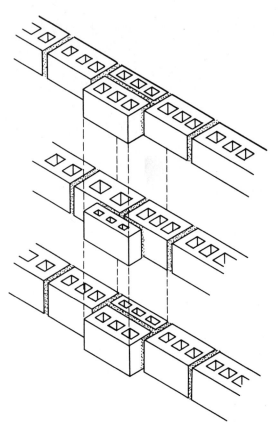

FIG. 16–27 Pilasters may be used to help strengthen a wall. Overall design of the structure plus earth conditions are factors in determining the need, size, and frequency of such reinforcement. Check local building codes. All pilaster units should have full mortar bedding.

WALL CAPS

Walls should be topped, if only for looks. In the case of foundation walls, topping is necessary so that loads from above will be more equally distributed. The top course of concrete block can be made solid by filling the cores with concrete. Prepare for this during construction by placing a strip of metal lath in the very last joint (the one that is directly under the top course). The lath forms a bottom to contain the concrete core fill. If the material is available, you can cap with special top blocks that have a solid, 4-in.-thick concrete cross section at the top.

If permitted by local codes, 4-in.-thick, solid masonry units can be used to top off concrete block foundation walls. These can be placed directly over open

FIG. 16–28 Sealing off the top course of a hollow block wall, whether for looks or strength, can be done as shown here. Include a strip of metal lath in the joint under the top course.

FIG. 16–29 The mesh or lath acts as a floor to contain the concrete or mortar that will be used to fill all the cores in the top course of block. Level off the fill with a trowel.

Fig. 16–30 If available, special top blocks may be used as the last course. These have a top web that is solid concrete, 4-inches thick. Place them as you would any other block.

Fig. 16–31 Sometimes it's possible to use 4-inch thick, solid masonry units as a topping. Strength factors are not so critical when the job is done for appearance—more important when the wall must support, for example, a house.

cores. Use a full mortar bed and be sure that all vertical joints are completely filled.

ANCHOR BOLTS

Anchor bolts in concrete block walls must be set in concrete-filled cores. The way to do this is to anticipate the placement and its position when you are two

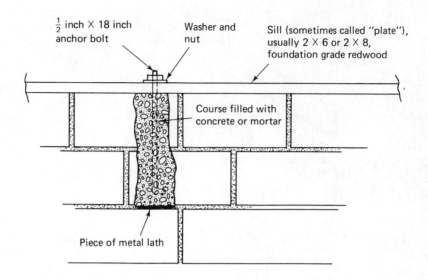

$\frac{1}{2}$ inch X 18 inch anchor bolt

Washer and nut

Sill (sometimes called "plate"), usually 2 X 6 or 2 X 8, foundation grade redwood

Course filled with concrete or mortar

Piece of metal lath

FIG. 16–32 How the sill is anchored to a concrete masonry wall.

courses from the top of the wall. At this point you can place suitably sized pieces of metal lath in the joint to support the concrete you will use to fill the cores when the wall is complete. Place the lathe so that cores can be filled to take anchor bolts on 4-ft centers.

CONTROL JOINTS

Control joints are necessary in masonry walls so that movements caused by various types of stresses can occur without damage to the structure. The control joint can be a continuous vertical joint designed into the wall where the stress will be concentrated. To form a continuous vertical joint, combine half-length and full-length blocks at that point.

One type of control joint can be built with stretcher block if noncorroding, Z-shaped, metal tiebars that are 2 in. narrower than the width of the wall are used in alternate horizontal joints to provide lateral support for wall sections on each side of the joint.

Another popular control joint is made by inserting roofing felt or building paper into the end core of the block. The paper should be long enough to match the full height of the joint and wide enough to extend across the joint. The paper prevents the mortar from bonding mating units. For lateral support you can fill the core with concrete or mortar.

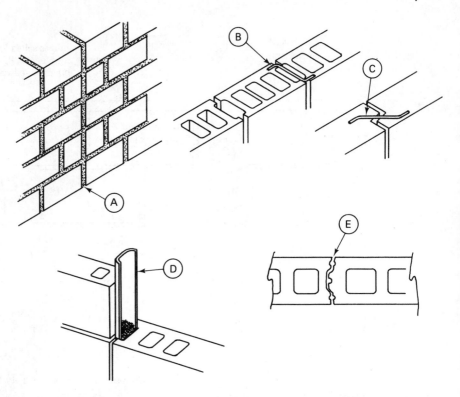

FIG. 16–33 Control units: (a) deliberate, continuous vertical joint, achieved by using half and full units; (b) the use of a metal tiebar will provide lateral support to wall sections on each side of the joint; (c) offset jamb blocks are often used at control joints—note the metal tiebar; (d) filling ends of control-joint blocks with mortar will provide strength while a continuous strip of heavy roofing paper prevents the bond; (e) lateral support may be gained by using special tongue-and-groove blocks; these are available in half and full units.

WATERPROOFING BASEMENT WALLS

The exterior surface of concrete masonry walls, especially those that are partially or wholly backfilled, should be covered with a portland cement plaster or the same mortar that is used for placing the block. The total thickness of the coating (called *parging*) should be about ½ in., but it is best done in two applications.

Be sure that the wall surface is clean and wet. Wetting can be done just before the plaster coating is applied. Don't soak the blocks: The point is to prevent the dry blocks from sucking too much moisture from the plaster.

Apply the plaster to the wall with a trowel. It does not have to be beautiful, especially if backfilling will follow. But do aim for full coverage of uniform thickness. Let this application harden a bit and then scratch its surface to help pro-

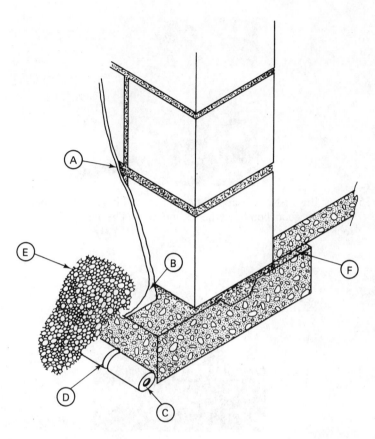

FIG. 16–34 Good basement wall design: (a) exterior of wall (earth side) is plastered (called "parging"); (b) end parge coat in cove to edge of footing; (c) place drain tiles; (d) cover drain tile joints with strips of felt or building paper; (e) cover to above footing surface with coarse gravel; (f) cast basement floor should rest on a bed of sand or felt.

FIG. 16-35 The concrete block should be damp (not soaked) when you apply the first parge coat. The coating may be mortar or a cement plaster that you can make by mixing one part of portland cement with 2½ parts of sand.

FIG. 16–36 Allow the first parge application to set a bit and then roughen its surface to help provide a good bond for the second coat. The mason here is using a special scratch tool.

FIG. 16–37 Allow the first parge coat to set for about 24 hours but do not permit it to dry out. Wetting can be done with a portable sprayer or with a garden hose. Frequent, light spraying is best.

FIG. 16–38 Apply the second parge coat as you did the first one. Work for full, uniform coverage. The total thickness of the parge coating should be about ½-inch.

FIG. 16–39 End up at the bottom (where the block meets the footing) with a generous cove that extends out to the edge of the footing. This will prevent water from accumulating at the joint line.

vide a good bond for the second coat. The scratching can be done with a special tool or with a whisk broom. The bristles of the whisk broom should be shortened and cut to provide open spaces. Keep the first coat damp while you allow it to set for about 24 hours.

Apply the second coat as you did the first but this time skip the scratching operation. Be sure that you form a generous cove at the bottom of the wall where the block meets the footing so as to prevent water from collecting at the joint line. Keep the second coat damp for about 48 hours.

PATCHING AND CLEANING

You can avoid many postconstruction chores if you avoid smearing mortar on the face of the blocks. When mortar droppings do stick, leave them until they

FIG. 16–40 Sometimes, it's possible to remove mortar stains by rubbing the spot with a broken piece of block. Work gently and don't over-do or you may create a blemish that is as bad as a mortar stain.

FIG. 16–41 A shop bench-brush can be used for some cleaning. It's okay to use on joint lines after they have been tooled and the burrs have been removed with a trowel. It will also remove dried materials that have not adhered to the block.

dry and then remove them with a trowel. Trying to clean up while the mortar is wet can smear some of the material into the face of the block and make it very difficult to remove. Check for any holes in the mortar joints and fill them immediately with fresh mortar. Don't use mortar that is too wet, however; the patch mortar should have the same consistency as the batches you made for laying the block.

HOW TO CUT BLOCKS

Whenever possible, design so that standard-sized units may be used to fill out a course. If you do need to cut a block, however, it may be done with the ham-

FIG. 16–42 You can work with a chisel and hammer. Hammer gently—score the block on opposite surfaces. Continue scoring until the break occurs.

Fig. 16–43 Special blades or abrasive discs are available for cutting masonry units with portable or stationary power tools. These do make clean cuts, but use them safely—wear safety goggles. Keep your hands and body clear of the cutting line.

mer and chisel used for cutting brick. You'll have to proceed carefully because of the block's hollow areas. Score the block on opposite surfaces by tapping the chisel gently with the hammer, repeating the scoring as many times as you have to to get a clean break.

Special masonry-cutting saws available for both stationary and portable power tools make very clean cuts but must be used carefully to avoid injury. It is also a good idea to protect your eyes with safety goggles.

QUESTIONS

16–1	Tell the difference between a full mortar bed and a partial one.
16–2	How do you start the first course of blocks?
16–3	Where do you place mortar on the blocks that follow?
16–4	State a good procedure for block work.
16–5	How do you gauge the alignment and the height of courses between the corners?
16–6	Is it possible that full mortar beds may be required in all courses?
16–7	Is an interlocking masonry bond necessary for an intersecting *bearing* wall?
16–8	What should be the design for an intersecting bearing wall when it does not occur at a corner?
16–9	How does the design differ from a *nonbearing* intersecting wall?
16–10	How do you provide for anchor bolts when you are constructing a block wall?
16–11	How far apart should anchor bolts be placed?

16–12   Why are control joints necessary in masonry walls?
16–13   Make sketches of two types of control joints.
16–14   What is *parging*?
16–15   How thick should a parge coat be?
16–16   How can block be cut?
16–17   What precaution is common to both methods?

17

Walks, Paths, and Driveways

Such projects as walks, paths, and driveways should be regarded as major factors in any design scheme—second in importance only to the house itself. These private roads and highways can keep people from trampling a lawn and can direct traffic as you wish it to move. Masonry underfoot helps to keep a house clean and will add to appearance and value regardless of whether the residence is in a city or rural area.

Good planning leads to projects that are durable and functional and that form an integral part of the overall landscaping theme. Ideally, the lot should be barren to start with so that designing doesn't have to work around existing projects. But even with a "used" house, which will have some degree of landscaping, the view for the future should be overall; existing projects that conflict with what you have in mind should be removed. Often, material salvaged from a razed project can be reused.

Think imaginatively. A walk doesn't have to be a straight ribbon of concrete, even though it might be cheaper to do than one with a gentle S-shape. Extras should be reserved for those areas that will stand out visually—for example, entry and leisure areas. You can be strictly practical in such hidden areas as along the sides of a house, behind a garage, and so on. Prosaic projects can be disguised further with such elements as screen block walls, which can play a more important role in the landscaping.

FIG. 17–1 You can view a walk or path as a series of separate pours. Large, temporary divider strips leave spaces that are then filled with soil and planted with grass.

Walkways should provide good traction, so a too-slick surface is out. The concrete is usually finished with a float or with a broom or by exposing the aggregate by washing. The choice among these and other finishes is up to you so long as it provides good footing.

Entry walks are normally 3 to 4 ft wide, service paths about 2 ft wide. These are guides only, however; you can do more or less in relation to available space and projected use.

Walks should be sloped about ¼ in. per foot of width to provide for drainage. The slope can be increased to as much as ½ in. per foot if conditions require it. Always slope away from the house and toward a general drainage area.

FIG. 17–2 Curves can add a lot of interest and can facilitate traffic flow since the pedestrian doesn't have to turn a sharp corner. Note the divider strips which also act as control joints.

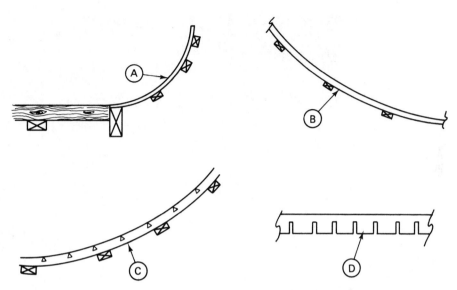

FIG. 17–3 To go from a straight line to a curve, you can work with ¼ inch plywood or hard-wood (a). One inch stock can be bent (b) if the curve is long and gentle. Kerfing (c) will permit bends in 2 x material. These are saw cuts (d) about three-quarters the thickness of the stock. The closer you space the kerfs, the more the board will bend. In all cases use as many stakes as you need to maintain the curve.

Keep surface levels on the same level as the general grade so that you don't form water pockets. When this is not possible or desirable, give extra thought to the drainage of adjacent areas. For example, if a lawn slopes toward a walk that is high enough to form a dam, it would be wise to bury drain tiles abutting and parallel to the walk to collect excess water and direct it to a drainage area.

FIG. 17–4 This interesting effect was achieved by making forms that produced round corners on separate slabs. Spaces between slabs can be planted in grass or you can do exposed aggregate "dividers."

FIG. 17–5 Changes in grade do not have to be done with steps. Often, a slight ramp provides more convenience for bicycles, lawn mowers, etc. Note the design grooves and changes in surface texture.

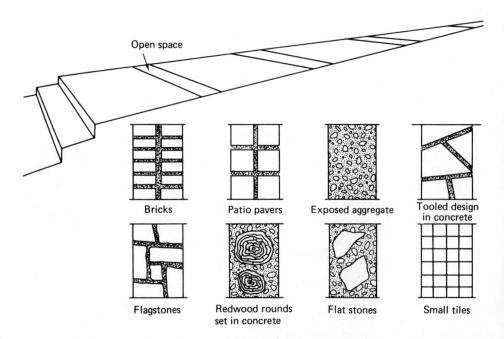

FIG. 17–6 To break up the monotony of long concrete walks leave open spaces to be filled with materials such as the ones shown here.

Walks do not have to be all concrete, or concrete at all. Such materials as brick, flagstones, patio tiles, and precast blocks can be used effectively, as well as stepping stones or even compacted redrock or loose gravel.

CONCRETE WALKS

Residential walks should be 4 in. thick and poured on subbases that have received the careful attention that we have already described. Use 2 × 4 material for sideforms but be aware that the net width is less than 4 in. Compensate for

Fig. 17–7 Stretch out a mason's line to indicate positions for sideforms and correct heights. Drive stakes deep enough so they will be firm. Extra care with forming will pay off later.

Fig. 17–8 Permanent sideforms or isolation joint material can be studded with nails so good bonding will occur with the pour. Use 16d galvanized nails for this purpose.

FIG. 17–9 Sand may be used to even out a subbase. It should be dampened and tamped so it will be firm. Projects like this should be planned to provide drainage away from the house.

FIG. 17–10 Follow the correct rules when dumping. Set concrete as close as possible to its final position. Do minimum spreading especially if you work with a rake. Use a square-edge shovel to settle material against the sideforms.

this by having a subbase surface that is a bit lower than the bottom edge of the forms. Take your time in preparing and bracing the formwork. The preliminary work is as important as the quality of the concrete and the craftsmanship of the pour and the finishing. The trade calls this kind of work "one-course construction," which means that the full thickness of the concrete is placed at one time. Exceptions to this occur when a special topping is used—for example, exposed aggregate—or when concrete serves as a base for brick, flagstone, or tile.

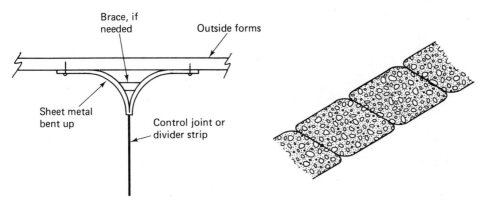

FIG. 17–11 Getting a round corner effect on path slabs.

JOINTS

The stability of concrete is affected by moisture and temperatures. It has the most volume immediately after being placed, the least when it has dried out. This normal expansion and contraction can result in cracks even in the best of work unless control measures are included in the design.

Control joints or *contraction joints* are cut into concrete to deliberately provide a weak line that will induce cracks, if they occur, to appear where you want them to. Ideally, the crack will happen under the control joint and will not be visible at all.

Walks (as well as driveways and patio slabs) should have control joints across them with spacing equal to the project's width. If the project is wider than 10 ft, a centered, longitudinal control joint is wise. Generally, spacing of control joints in any concrete slab project should not exceed 10 ft in any direction, and the panels created by the joints should be approximately square. Cracks are more likely to occur when there is an excessive length-to-width ratio.

Groovers or *jointers* are used to cut control joints. They are designed to cut the concrete to the correct depth and to form a neat radius on either side of the cut. Tools similar to these that do not cut as deep are often described as *cheaters*, since they are used for decorative purposes but match the appearance of control joints.

Use a straight piece of wood as a guide for the groovers. Start the pass by pushing the tool into the concrete and then moving it along the board. Put pressure on the *back* edge of the tool. To get a smooth finish, repeat the procedure but move the tool in the opposite direction. If the concrete has set enough to make the grooving operation difficult, you can work by tapping the groover with a hammer to form the basic line, then finish with the normal passwork.

Control joints are not required on gridded slabs. Also, you can use permanent, thin strips of wood in place of the cut joints (see sketch).

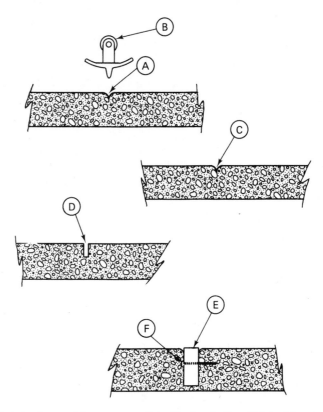

FIG. 17–12 Control joint (a), done with a groover (b). Its depth should be at least one-fifth the slab thickness and the cut should be slightly rounded (c). The joint may be cut with a masonry saw (d) to at least one-fifth the slab thickness but no wider than ¼ inch. 2 x 4 wood divider strip (e) may be used. Add 16d galvanized nails (f) spaced about 16 inches and driven from alternate sides.

Isolation joints (sometimes called *expansion joints*) are required to ease potential stresses where the slab abuts an existing walk, driveway, or house, and rigid objects such as clothes or utility poles. Premolded strips of fiber material are usually used for this job. The thickness of the material can be from ¼ to ½ in., but its width should match or slightly exceed the thickness of the slab. Be careful when installing isolation joints in traffic areas. The joint material should never project above the slab's surface; it is even a good idea to place it slightly below the surface to eliminate a potential tripping hazard.

A *construction joint* is used when the pour is done in sections and no permanent divider strips are included in the design. Professionals prefer to avoid a construction joint, but when the scope of the work or the time available makes it necessary, it should be located so that it will act as a control joint. The purpose of the construction joint is to provide a keyway so that abutting slabs will interlock and work together to keep a level surface.

FIG. 17–13 Never cut a control joint without using a straight board as a guide. Keep pressure down on the heel of the tool. Two passes, in opposite directions, are usually enough.

FIG. 17–14 Edges are formed in a neat radius with an edging tool. Keep pressure on the heel of the tool and on the edge being shaped. This operation is often skipped when the formboards are permanent.

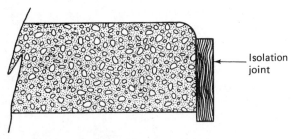

FIG. 17–15 Use *isolation joints* between the pour and existing buildings or walks or curbs. The joint is a special, premolded material that is ¼ inch to ½ inch thick. Keep it slightly below slab surface (¼ inch) to avoid a tripping hazard. Okay to install flush, for example, where the slab abuts a building.

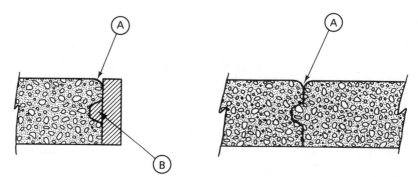

FIG. 17–16 *Construction joints* (a) are edged to match the control joints. The key (b) for slabs 4 inches to 5 inches thick can be made by beveling a strip of 1 x 2.

Special keyway materials are available for the job, but nailing a beveled 1 × 2 to a piece of 2 × 4, as shown in the sketch, will work as well. The top edges of the construction joint should be tooled with a groover.

BRICK WALKS

Brick walks can be placed on a concrete subbase or on a bed of sand. In either case, normal-sized bricks or special pavers can be used, depending on whether you wish to use joints in the design. When a concrete subbase is used, the bricks are set in a bed of mortar—much like building a wall flat on the ground.

FIG. 17–17 Brick walks can be done so as to blend with any landscaping scheme. As shown here, you can introduce curves for gentle lines. Bricks are most permanent when done over a concrete subbase.

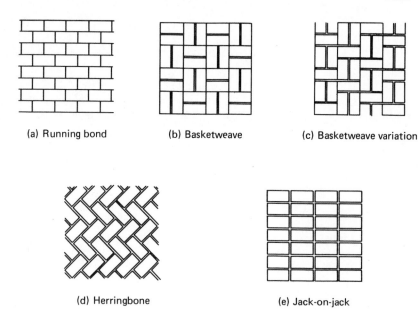

(a) Running bond   (b) Basketweave   (c) Basketweave variation

(d) Herringbone   (e) Jack-on-jack

FIG. 17–18 Suggested pattern bonds for brick walks—or patios: (a) Running bond is a straight-forward pattern that requires very little cutting. Plan the width of the project to match the total width of the bricks; (b) Basketweave requires accurate laying but the project may be planned to eliminate cutting entirely; (c) Basketweave variation is often called "half basketweave." It *does* require cutting; (d) Herringbone is widely used but one of the more difficult to do. It calls for a lot of accuracy and much cutting; (e) Jack-on-jack has continuous joint lines in both directions. Uniform joint width is essential but the project may be planned to eliminate cutting.

Prepare the site as you would for a concrete walk but leave top space for the thickness of the brick plus the thickness of the mortar bed. Subbase preparation is equally important when you plan to place bricks on sand. In this case, though, you build up just enough to allow for the thickness of the brick.

You can use standard pattern bonds or invent your own. Do give ample consideration to the amount of brick cutting you must do to achieve a particular effect. Less work and time are required when the width, length, and pattern bond of the project are decided with minimum cutting as an objective.

It is a good idea to use sideboards as part of the project. Perimeter bricks can be loosened by traffic, the weight of a lawnmower, and the like; sideboards will provide support.

Sand subbases should be leveled, wetted, and tamped. The need for firmness is obvious. The better the job, the more stable the brick. The sand bed may be built up slightly higher than absolutely necessary. Then the bricks may be tamped down with a rubber mallet so that they will be firmly embedded. Don't overdo the extra sand-bed height; it shouldn't be necessary to pound with the rubber mallet as if you were using a sledge.

Mix the mortar for on-concrete work as you would for normal bricklaying.

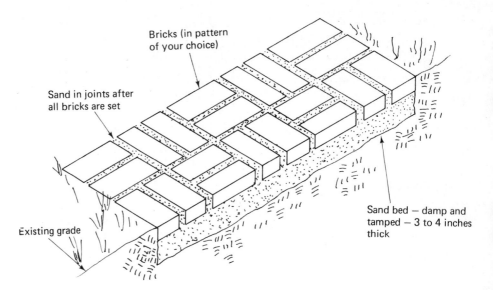

FIG. 17–19 Brick walk on a sand bed.

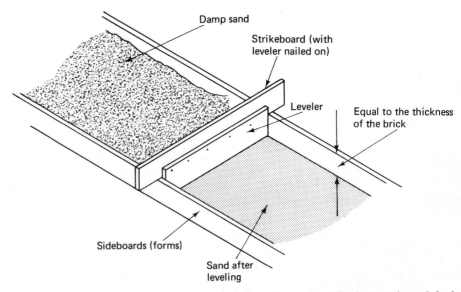

FIG. 17–20 How to level a sand bed for a brick-on-sand walk. NOTE: Thickness of sand bed should be 3 to 4 inches over well-compacted base.

The mortar bed should be fairly level and a bit thicker than necessary so that you can press the bricks down firmly to seat them.

If you intend to use joints, be careful with spacing. Have a half dozen small pieces of plywood of correct thickness on hand to use as gauges. These can be reused from spot to spot as you progress with the work. Use a straight piece of

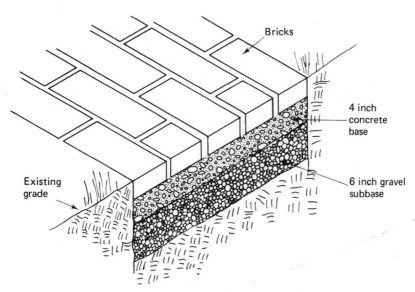

Bricks

4 inch concrete base

6 inch gravel subbase

Existing grade

F IG . 17–21 Local codes may dictate this design for a brick walk or patio. In any event, it is a good heavy-duty, brick-on-concrete design. The bricks are laid in, and the joints are filled with mortar.

2 × 4 to check for overall levelness as you go. The 2 × 4 should be about 4 ft long so that it can span quite a few bricks and so that it may be used as a tamper to achieve uniform height.

Joints may be done in several ways. You can simply spread sand over the project and then work with a broom to fill the spaces. Use a fine spray from a garden hose to settle the sand and then, if necessary, repeat the procedure.

You can elect to mortar the joints in normal fashion or do a faster, less tedious job as follows: Mix the mortar ingredients in routine fashion but do not add water. Spread the dry mix over the project and work with a broom to fill all the spaces. Continue this until you are sure that the joints are full and level. Wet the project down with a garden hose, using a very fine spray and pointing the hose upward so the water will fall like a gentle rain.

Paver bricks abut without joints, but here, too, it is not a bad idea to go through the sand–broom–wetting procedure that we described previously.

OTHER MATERIALS

Flagstones or patio tiles are often used for walks. Either of these materials may be placed on sand or on a concrete subbase. Site and subbase preparations are the same as those outlined for brick. Patio tiles (or pavers) are uniform in area size and thickness, so placing them calls for fairly standard procedures in relation to bedding, spacing, and mortaring. Flagstones are purchased in large pieces

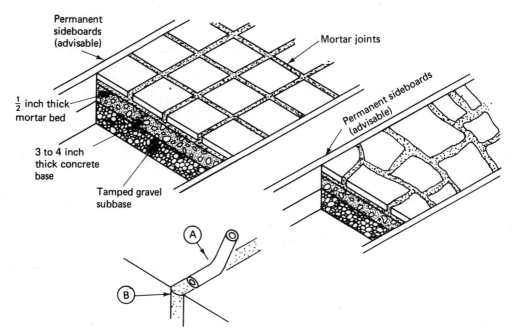

FIG. 17–22 Tile or flagstone walks. While both these materials (like brick) may be done mortar-less on sand, a long-lived installation that requires minimum maintenance should be done this way. Do the joints with a concave tool or a bent-up piece of pipe (a) to get a slightly concave shape (b).

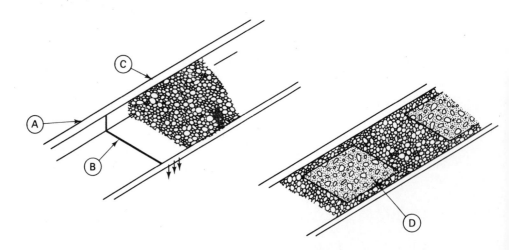

FIG. 17–23 Loose gravel walk: (a) use permanent sideboards to retain the gravel; (b) building paper or sheet plastic underlayment; (c) tamp gravel level with top edge of side-boards; (d) island stepping stones may be placed with the gravel.

that often vary in thickness. This calls for cutting and for adjustments in the bed.

An irregular pattern bond is almost always used for flagstones, but if this isn't planned in advance with some regard for the overall appearance, the effect may be lost. It is smart to work on paper, planning a pattern bond that will be repeated every 3 or 4 ft. It even pays to make paper templates for each of the shapes in the bond so that you can use them as patterns for marking the stone. You can cut curves and round corners, but it is best to design with straight lines.

Cutting isn't difficult and is accomplished with a brick chisel and hammer. Place the stone on a firm flat surface and score the cut line with the chisel. Raise the stone on a couple of 2 × 4's with the outer edge of one of the platforms just inside the score line; then snap off the piece to be removed. This can often be accomplished with your own weight. Keep one foot on the "good" side, and bear down with your other foot on the piece of stone that must be broken off. Irregularities can be smoothed out by working with the hammer only. Wear safety goggles.

DRIVEWAYS

Structurally, concrete driveways are just oversized walks. A full 4-in. thickness is enough if you limit traffic to passenger vehicles, but it makes sense to go to 5 or 6 in. if only to take care of that occasional truck. Widths for a single car run from 10 to 14 ft but 14 ft is a minimum if a curve is involved. A general rule, when space permits, is to make the driveway about 3 ft wider than it must be to accommodate the car. If there is a long approach to a two-car garage, the entry portion of the driveway may be single-car width, but it should be widened at the garage so that a car can get into either parking space. When the driveway for a two-car garage must be short, then consider 16 ft as a minimum width, 24 ft as nice to have. Driveway design must, of course, be within the limits imposed by the available space, but good planning often makes it possible to include offsets and parking places even on small lots.

Changes in grade from street to driveway should be gentle and as gradual as possible to avoid scraping the undersides and bumpers of automobiles. The recommended maximum vertical rise per foot of horizontal run is 1¾ in. The driveway itself should be sloped enough so that rain water or water from a hose will drain off quickly. A slope of ¼ in. per lineal foot is the recommendation. Ideally, the slope should be toward the street. If conditions dictate otherwise, drainage can be accomplished with a cross slope or by shaping the pour as a raised or inverted crown. When crowning, the strikeboard itself is shaped to produce the arc.

Be sure to follow all the correct procedures that will assure a good project.

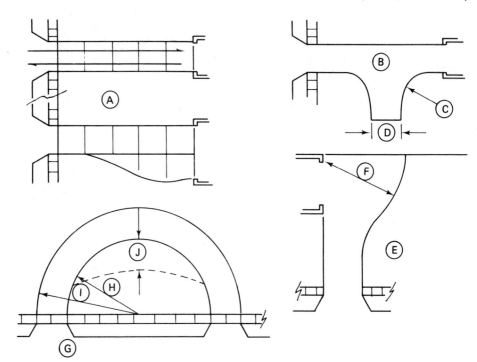

Fig. 17–24 Thoughts on driveway design. (a) Straight driveways are adequate but mean backing out into the street; (b) it's better if you can provide a turning area; (c) about an 18 foot radius; (d) width of driveway; (e) it can also be done this way; (f) 35 to 40 feet; (g) circular driveway is most convenient; (h) about 18 feet; (i) about 31 feet; (j) if you want a passing area or parking area add 10 feet here.

These include layout, site preparation, formwork, and so on. Although the usual finish for a concrete driveway is done with a float or a broom, there is no reason why you can't do one of the more exotic finishes already described.

Do remember that a driveway usually covers a large area, and that it is possible for some of it to do double duty. When planning, think of such additional uses as a post for a basketball backstop, a net for volley ball, perhaps even a shuffleboard court.

HOW TO ESTIMATE CONCRETE NEEDS FOR DRIVEWAYS*

Thickness of Slab (in.)	Number of Square Feet					
	10	25	50	100	200	300
4	0.12	0.31	0.62	1.23	2.47	3.70
5	0.15	0.39	0.77	1.54	3.09	4.63

*Figures are cubic inches. To find amounts for areas not shown, total the pertinent columns. For example, for a 375 ft.² project, 4 in. thick, add 3.70 + 0.62 + 0.31. In all cases, add 5–10 percent allowance for uneven subgrade and spillage.

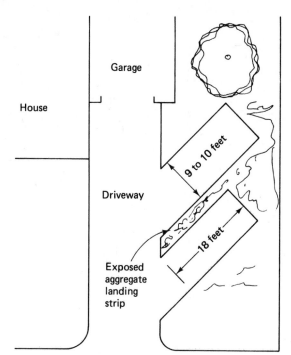

Fɪɢ. 17–25 Off-driveway parking spaces.

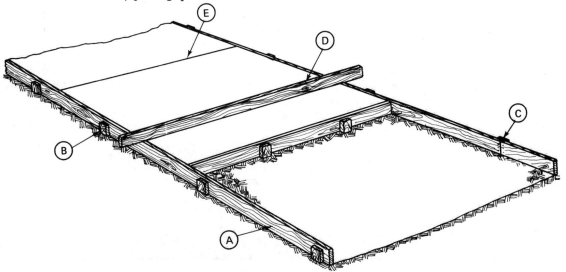

Fɪɢ. 17–26 Guides for a driveway pour: In general, use a 10 foot minimum width—slab-thickness minimum (4 inches), but 6 inches is better—pour concrete on well-tamped, firm earth or a gravel subbase—pitch (across) for water run-off should be minimum of ⅛ inch per foot, but ¼ inch is better. Use 2 x 4 sideboards (a) well-staked at frequent intervals (b). Back up splices of sideboards with stakes (c). Use 2 × 4 riding edges of sideboards to level concrete after the pour (d). Cut control joints every 10 or 15 feet (e). Control joint depth should be one-fifth to one-fourth the thickness of the slab.

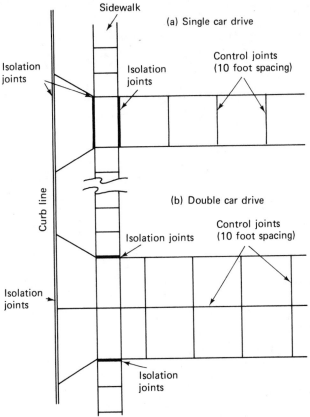

FIG. 17–27 Places for control and isolation joints in driveways: *isolation joints* in (a) are shown for drive installed to existing sidewalk; isolation joints in (b) are shown for sidewalk installed after the driveway.

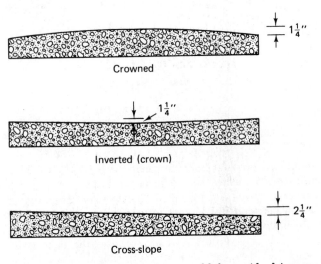

FIG. 17–28 Driveway drainage—driveway allowances assume a 10 foot wide driveway.

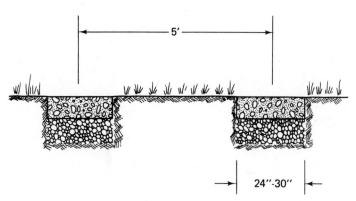

Fig. 17–29 Two, 6 inch thick concrete "ribbons" can serve as a driveway. Area between can be lawn, compacted gravel, brick-on-sand, etc.

QUESTIONS

17–1   What is an important consideration when designing?
17–2   How wide should walks be?
17–3   How much slope should you provide for drainage?
17–4   What is a good thickness for residential walks?
17–5   What is a *control joint*?
17–6   What does an *isolation joint* do?
17–7   What is a *construction joint*?
17–8   Make cross-section sketches to show a control joint, an isolation joint, and a construction joint.
17–9   What are two basic methods of putting down a brick walk?
17–10   Make a cross-section sketch of a brick walk that is placed over a concrete sub-base.
17–11   Make a cross-section sketch of a brick walk that is placed over a sand bed.
17–12   Describe, briefly, three methods of filling joints in a brick walk.
17–13   Name two materials that may be used for walks in place of concrete or brick.
17–14   How thick should a driveway slab be?
17–15   What is a minimum width for a driveway?
17–16   What if the driveway is curved?
17–17   What is a good general rule to follow when designing a driveway?
17–18   Make several sketches to show how a driveway can be designed to provide turn-arounds and parking spaces.
17–19   What is an important consideration in relation to changes in grade from drive-way to street?
17–20   What other uses for a driveway can you think of that you should keep in mind when first designing the driveway?

18

Stepping Stones

Stepping stones are popular because they can serve many functions, will fit any landscaping decor, and are fun to do. A single stone may be used as a step across an open area or through a flower bed, or a series of them may be set down in place of a walk to provide clean footing for occasional traffic. Often, large units are set down as a walk, in a straight or curved pattern, because they tend to scale down the size of the project. This may make the project more appropriate for the setting or for the size of the lot.

Stepping stones can be more attractive than concrete ribbons and much easier to do since they may be cast as modular or irregular units as time and energy allow. The craftsman may be very creative with shapes, textures, color, and embedded materials. Common shapes are round or square but they can be elliptical, octagonal, rectangular, free-formed—even star shapes or giant footprints. Surface textures can be any of those described for concrete finishing— exposed aggregate is very popular—but you can go far beyond the conventional effects to produce practical and unique projects.

The thickness of units can range from 1½ to a full 4 in. Base your choice on the size of the stone, its placement, and the degree of duty. For example, large, cast-in-place units that form a main walkway should be 4 in. thick and placed with all the attention you would give a full pour. A 2-in. thickness isn't

FIG. 18-1   Large stepping stones are often used in place of a full concrete walk. The stones may be cast in place or as units that are then transported to the site.

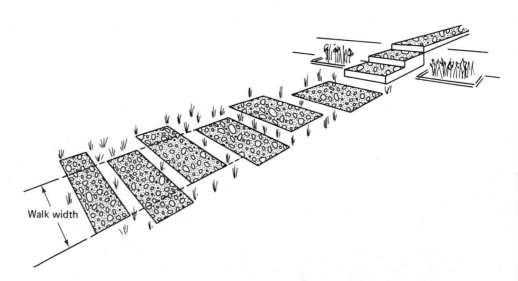

Walk width

FIG. 18-2   A more interesting effect is achieved when large stepping stones, used as a walk, are staggered. The actual walk width is the amount the stones overlap.

bad for casual stones. They will be economical to make and not too heavy to move from forming site to the placement area.

A good concrete mix would be 1–2–2¼, with maximum aggregate size in line with thickness of the stone. When using mortar, entirely or as a topping, stay with the proportions called out for heavy-duty applications.

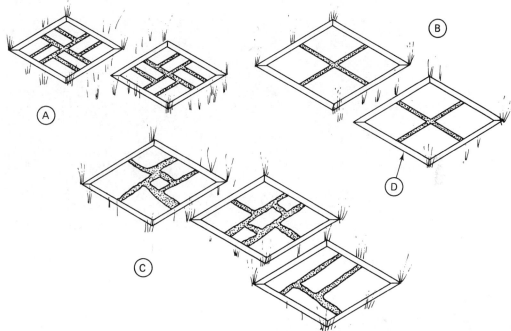

Fig. 18–3 For something different in the way of stepping stones, you can work with bricks (a),
tiles (b), or pieces of flagstone (c). In each case, the job will look better longer if
you use wooden frames; these may be preassembled. NOTE: Doing such work directly
on soil is not a good idea—using a sand bed is acceptable, but laying on a concrete
bed is best.

Fig. 18–4 These are examples of ready-made stepping stones that are available at building supply or garden supply establishments. Here, they are used to supply clean footing
across an open area.

FIG. 18–5 Being carefully careless with stepping stones can supply a rustic project that does the job and fits nicely outdoors. These were cast with concrete that was left over from a major pour.

FORMS YOU CAN MAKE

Straight-sided stepping stones that are made above ground are usually cast in simple wooden forms. The forms can be made for a single stone or for multiple casting; they can be reusable or for one-time application only. The sketches show several types of knock-down forms as well as lift-up types. The latter are permanently assembled but are also reusable. The lumber you use depends on the thickness of the stone and the overall size. Common materials include 1 × 3's, 1 × 4's and 2 × 4's. Lumber of fairly good quality should be used for forming. If you work with low grades, you'll have trouble with loose knots, checks, splits, and warpage.

Sheet metal can be used with wood if you wish to design round corners or curved sides, or it can be used alone for round shapes. The type of corrugated sheet metal that is sold as lawn edging is nice for round stones, since it produces projects with a distinctive shape.

Forms for stepping stones do not require a bottom unless you wish to use a particular material to achieve a surface texture. For example, wire-brushed plywood will give the surface of the stone a three-dimensional woodgrain look. Normally, you can set the form down on any flat surface. Be aware that either surface of the casting can be the top of the stone. For normal finishing you can work on the exposed surface of the pour while the material is still in the form. When the bottom of the pour is to be the top of the stone, all you have to do is strike off the top surface. For smooth surfaces, you can set the form down on a sheet of hardboard or building paper. If you want a rougher texture, set the form down on sand or soil that is not too smooth.

When you pour, tamp only as much as you have to to settle the mix. Re-

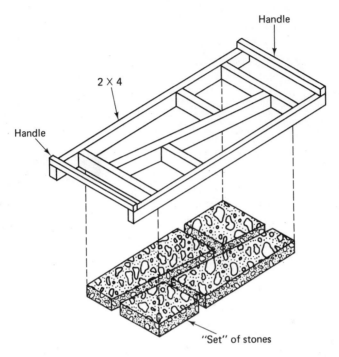

FIG. 18–6 Make special reusable forms when you wish to cast stones in a repeat pattern. Use 2 x 4 stock to make the form and attach a strip at each end as a handle for lifting the form.

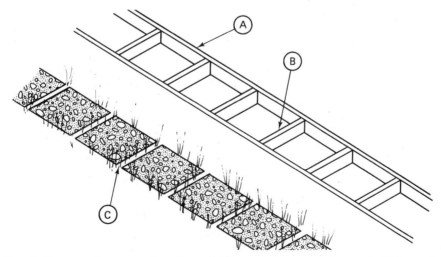

FIG. 18–7 Cast in-line stepping stones by using a ladder-type form: (a) 1 x material here okay; (b) use thin or thick material here depending on how much spacing you want between the stones; (c) fill between the stones with soil and then plant grass seed. NOTE: Keep stones slightly below level of lawn to avoid interference with lawn mower.

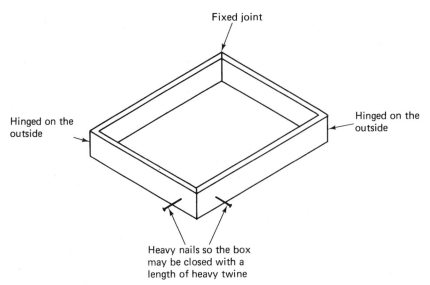

FIG. 18–8   Example of a reusable form for stepping stones. Use 1 x 4 material to make a bottomless box.

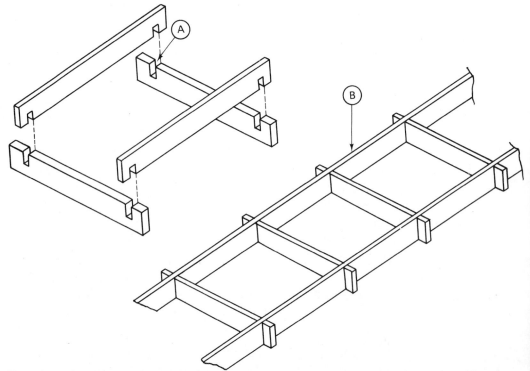

FIG. 18–9   Another way to make reusable forms for stepping stones: (a) 1 x material is notched to interlock—width of cut equals the thickness of the material; the depth of cut equals one-half the width of the material; (b) use the same idea to make a form for multiple casting.

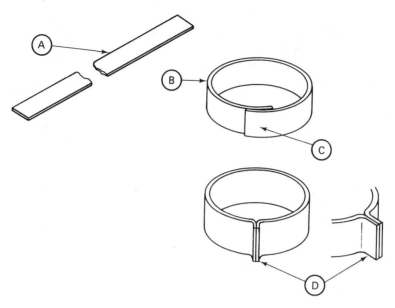

FIG. 18–10 One way to make a reusable form for round stones: (a) cut a strip of flashing (or similar material) about 3½ inches wide; (b) bend around suitable form such as ash can, oil drum, whatever; (c) overlap ends; (d) bend out ends to form flange that may be clamped tight.

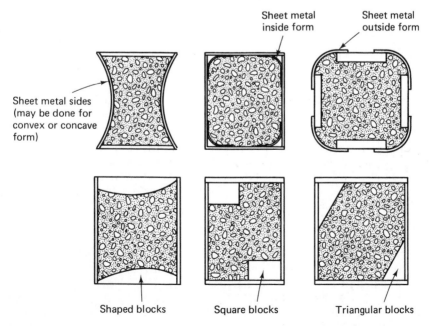

FIG. 18–11 Examples of how to make forms for stepping stones with unique shapes.

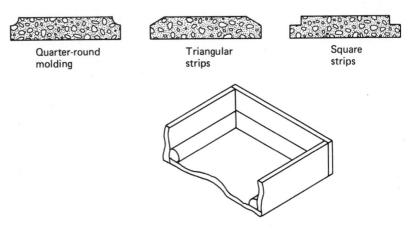

Quarter-round
molding

Triangular
strips

Square
strips

FIG. 18–12 Get fancy edges on stepping stones by using quarter-round molding, triangular strips, and square strips, which can be placed in the form as shown.

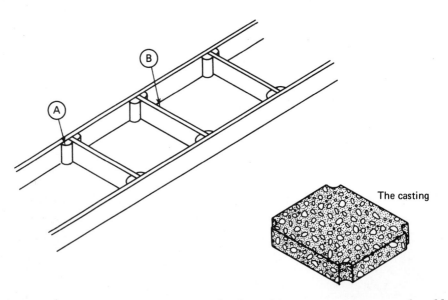

FIG. 18–13 Stepping stones with shaped corners can be done this way: use quarter-round molding (a) on side forms so that dividers (b) can be a slip-fit between them. NOTE: In place of the quarter-round molding you can use triangular or square strips—even picture frame moldings.

move the forms as soon as you can do so without damage to the pour. If you are very careful about correct water content of the mix, it will be possible to remove forms rather quickly, although the time lapse also relates to the temperature. At any rate, don't rush the procedure just because you want the forms for another pour. Should edge damage occur, repair the damage right away with a trowel.

Forms will be easier to remove if you apply a light film of oil to the contact surfaces or if you use a release agent. Stepping stones should be water-cured just as if they were a full concrete project.

FORMING IN THE SOIL

Forming in the soil itself is done mostly for casting in place, but there is no reason why the technique can't be used if you wish to avoid making wooden forms. An ideal application for in-place casting would be the establishment of a stepping-stone path across an existing lawn. All you have to do is make neat excavations to suit the size, shape, and thickness of the stones. You will get clean edges if you make perimeter cuts first, working with the edge of a square-bladed shovel as a cutting tool. Following this, if you wish, you can make a clean sub-

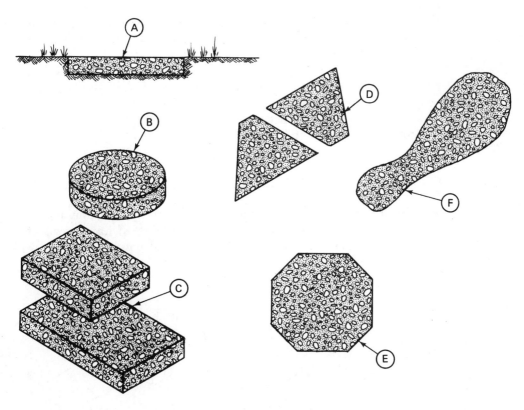

FIG. 18–14 You can dig in the earth to make forms for stepping stones. This is especially good if you can cast the stone where it will be used. Finish the surface as you would any slab—with float or trowel (not too smooth!) or by washing to expose aggregate (a). Projects may be round (b), square or rectangular (c), triangular (d), multi-sided (e), or you can do unique things like giant footprints (f).

surface cut that will remove the grass in small blankets that can be planted elsewhere.

Chances are that the subgrade you expose will be firm enough and damp enough to take the concrete right off. If not, tamp it firm and add sand to bring it to correct level. Work so that the level of the pour will be just low enough to prevent interference with lawn mowing.

A sand pile may also be used for nonwood forming. Level it and wet it enough so that it will hold the shape you want. Pour carefully so that you don't break down the walls of the form.

READY-MADE FORMS

Cardboard boxes are the most easily available ready-made forms. Seal them shut with gummed tape and then slice off sections that will produce the stone-thickness you want. Sides may bulge but you can use this to advantage if you wish to include a curve, or you can brace them with a concrete block or a bulwark of soil or sand. Such forms are for one use only. Peel them off as soon as the concrete sets and discard them.

Any item that can be used in a similar fashion may be utilized as a form for stepping stones. This includes, hoops, sections of plastic containers, garbage pail covers, parts of wooden crates, and the like. A good many of them are reusable. Just be sure that whatever you use is not undercut in any area so that it becomes permanently attached to the pour.

SURFACE PATTERNS

When you cast a stone with a mortar mix or with a mortar mix topping, you can impress it with some rather fine detail. Imprints of leaves, coarse fabrics, embossed medallions, wood cuts, and so on are all possible. You can work by

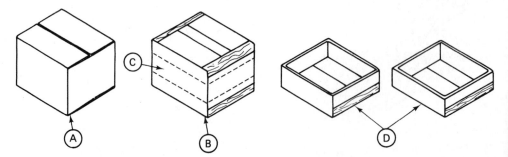

Fig. 18–15 A heavy cardboard box (a) can be closed tightly with tape (b) and cut on the dotted lines (c) to provide two one-time forms for stepping stones (d). Cut so the forms will be about 3 inches deep.

Fig. 18–16 An example of the fine detail you can impress in the surface of a stepping stone. This can be done if you work with a straightforward mortar mix or with a mortar mix topping. This design was done with a section from a fine, wrought iron screen.

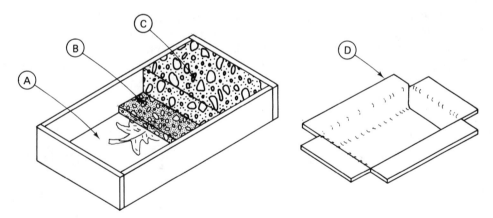

Fig. 18–17 For a very fine imprint, line the bottom of the form with sheet plastic (a); cover imprint material with ½ inch layer of mortar mix—let it set a bit (b); and then finish pour with concrete mix (c). You can accomplish this without a form box by lining an excavation in sand or soil with sheet plastic.

putting the imprint material in the bottom of the form and pouring over it, or you can pour first and then press the object into the surface when the material has set a bit. When you use the former technique, be sure to coat the decorating item with oil or a release agent.

Other surface patterns are possible by placing foreign objects in the bottom of the form before the pour so that they become a permanent part of the stone.

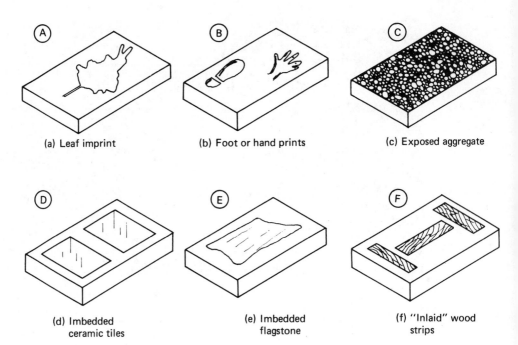

(a) Leaf imprint   (b) Foot or hand prints   (c) Exposed aggregate

(d) Imbedded   (e) Imbedded   (f) "Inlaid" wood
ceramic tiles   flagstone   strips

FIG. 18–18 Methods of surface-decorating stepping stones. A fine mortar mix will pick up all of the materials shown. With (a), (d), (e), and (f) place the objects in the bottom of the mold before making the pour. With (b) press down firmly with shoe or hand before the concrete sets. With (c) hose off surface to expose coarse aggregate, or tamp in special aggregate and then hose off before the concrete sets.

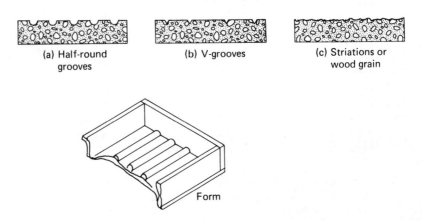

(a) Half-round   (b) V-grooves   (c) Striations or
grooves   wood grain

Form

FIG. 18–19 Shaped surfaces for stepping stones. For (a) and (b) place half-round or triangular strips in form. For (c) use striated plywood or hardboard or wire-brushed plywood as the bottom of the form.

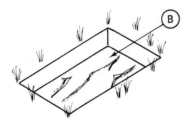 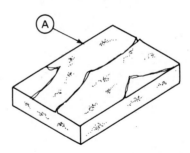

Fig. 18–20 When you use soil or sand as a mold, you can achieve natural stone effects (a) by texturing the bottom of the form (b).

You can use small tiles, pieces of flagstone, blocks of wood, flat stones, shells, and the like. Other ideas include footprints, handprints, house numbers, and so on. You can make a different kind of surface pattern by adding a bottom to the form. This, in itself, depending on the material you use, can supply a surface pattern; or you can attach longitudinal or lateral strips of shaped wood such as ½-round molding.

QUESTIONS

18–1	How thick should stepping stones be?
18–2	What are three methods of casting stepping stones?
18–3	How can you achieve a scalloped edge on a stepping stone?
18–4	Make a sketch that shows how a reusable form can be made.
18–5	What is a typical situation where casting-in-the-ground might be in order?
18–6	How can you get fine detail in the surface of a stepping stone?
18–7	Name some materials or objects that can be used as ready-made forms.
18–8	Design an unusual stepping stone. Make a sketch that shows the form required to cast it.

19

Floors for Outdoor Living

A great deal of advance planning is the key to a successful patio project. This applies as much to the design and projected use as it does to the actual construction. View the patio as the family's outdoor living space, and as a pleasant setting for entertaining guests. It should be as roomy as possible, certainly as spacious as the largest room in the house. Consider that you will spend much leisure time there, that you may use it for sunbathing, cooking, eating, listening to music, reading, even watching television.

All these factors dictate that the location and the design be affected by such things as climate, closeness of neighbors, view, traffic patterns, landscaping schemes, even the need, if necessary, to provide protection from insects. Existing situations may force a location that is less than ideal but you can add elements to compensate. For example masonry or shrubbery screens can be included to block a prevailing breeze if it is annoying or to provide privacy. Open or closed screens will also help to keep noise out, or in.

Protection from rain may call for an overhead structure; protection from insects may be achieved by including wire-screen walls. In each case, provisions for posts or framing may be included when the floor is placed. Footings for privacy or seat walls, posts for lights, underground electrical cable, pipes for water, legs for wooden benches, open areas for on-patio plantings—all these needs should be anticipated and planned for. It doesn't matter if they will be done immediately or at some future date. Providing for them will make future work easier as well as helping to avoid later remodeling.

Consider the climate when planning the patio exposure. A northern exposure will get minimal direct sun. A patio that faces south gets the most sun, whereas one that faces west will probably be cool in the morning and hot in the afternoon. Easterly exposures usually cool off in the afternoon. Consider these facts in relation to projected use.

Consider your gardening interests. Do you want a perimeter planter wall with flowers that require attention, or a border of minimum-maintenance, ever-

FIG. 19–1   View the patio or terrace or atrium—whatever you wish to call it—as an outdoor room that increases usable floor space. Plan in advance for all conveniences and pleasures.

FIG. 19–2   You have a broad choice of materials that range from poured concrete through brick and masonry units to precast slabs. Consider that deep joints, with any material, will make the pavement harder to sweep.

FIG. 19–3 Plan for components that will be added. Think of buried cable for electricity to lights or a barbecue area, and pipe for water. These materials are easy to install *before* the pavement goes down.

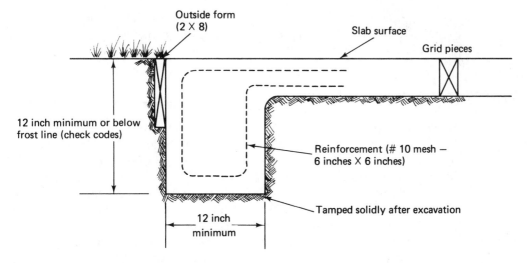

Outside form
(2 X 8)

Slab surface

Grid pieces

12 inch minimum or below
frost line (check codes)

Reinforcement (# 10 mesh —
6 inches X 6 inches)

Tamped solidly after excavation

12 inch
minimum

FIG. 19–4 If you intend to have a perimeter wall on the slab, the wall footing may be poured as part of the slab.

green shrubs? If you include or plan to add a barbecue later, proximity to the kitchen may be important. It is both useful and pleasant if the patio can be reached through large, sliding, glass doors from an adjacent family or living room. This tends to combine indoor and outdoor leisure areas and makes both seem more spacious, especially when the floors are on the same level.

Straight-sided patios are okay and much in evidence but the shape, really, is limited only by your imagination and is not dependent on the flooring mate-

Fɪɢ. 19–5 Gentle curves can be accomplished with 1 x material. Doubled ½-inch material is better when the curve is extreme. Note here that perimeter walls were done first.

rial you select. Curves for cast concrete call for more forming work; bricks, flag-stones, and precast units require cutting to conform to curved lines, but a little more effort can be amply rewarded. You and the project do have to live together.

THE MATERIALS

Any of the materials we have already discussed, including some types of pseudo-brick, can be used for the outdoor floor. Concrete is justifiably popular because it is durable and easy to cast in special shapes. Brick is often chosen because the effect can be "woodsy" and, in some situations, is cooler than other materials. Flagstones laid with precise joint lines can be quite formal. Some types of pre-cast units make the job easier because they are available ready-made and are often set down on comparatively easy-to-do sand beds.

All the structural and application considerations discussed previously apply to patio work regardless of the flooring material selected.

CONCRETE-PATIO GRIDWORK

A concrete patio may be poured as a full slab, but usually, for esthetic and even practical reasons, a system of permanent grids is included. A patterned slab makes it easy to leave open places for on-patio plantings. The grid system, in squares or rectangles, breaks up the driveway or parking lot appearance of a large slab and provides the opportunity to include a special design. The design de-pends on how much formwork you want to do and, of course, the effect you

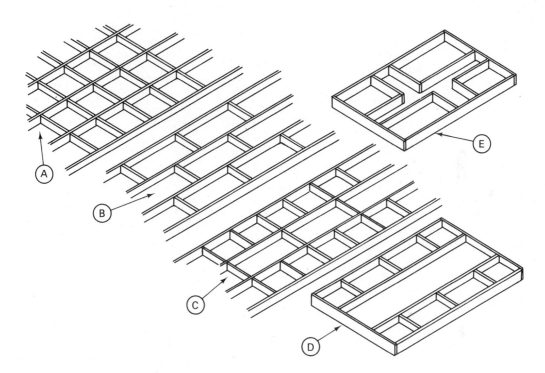

Fig. 19–6 Designs for gridwork: (a) in line—squares—may be rectangles; (b) staggered—may be squares or rectangles; (c) combination of (a) and (b); (d) and area without divisions (center) can be used to direct traffic; (e) almost any pattern is possible— thus one might be suitable for a small slab.

want: Radiating web lines, geometrical patterns, circles, diamonds, and the like —all are possible.

From a practical sense, a gridwork patio may be viewed as a series of small slabs, to be filled in as time and energy permit. The slabs will have an overall uniformity so long as you are careful with mix proportions and the finishing. Gridwork forming also makes it possible to do the job with different materials. For example, you might alternate concrete squares with brick squares.

Heart redwood, because it is highly resistant to decay and insect damage, has come to be accepted as *the* material for concrete gridwork. Since all you'll see of the material after the concrete is poured, is the top edge, it is all right to use an economical grade, but don't be too casual in the selection. Pieces that twist or wind or are overly warped can be impossible or, at least, a nuisance to install. Some workers will buy a quantity of cheap material and cull out the good stuff, but this is a risk that may end up costing more money, and certainly more time.

Some workers will cut 2 × 4's in half lengthwise and use the resultant 2 × 2's as grid pieces. This saves some money but it doesn't help produce

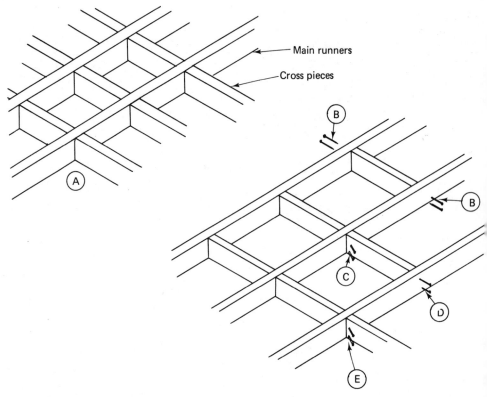

FIG. 19–7 When grid crosspieces are in line (a), start by end-nailing the first piece (b). Add second piece by toenailing one end (c) and end-nailing the other end (d). Continue so (e) across all the main runners.

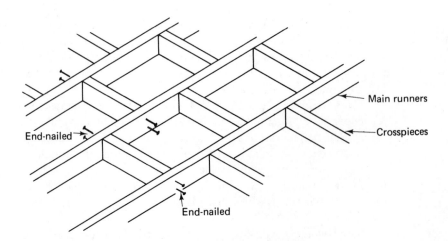

FIG. 19–8 When the crosspieces are staggered, all the parts may be end-nailed.

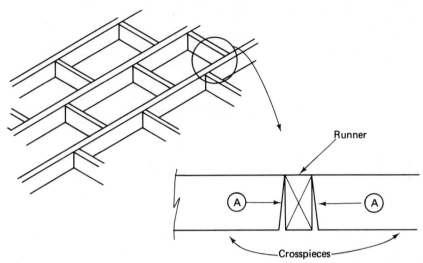

FIG. 19–9   For a tighter joint at the top (where it is important) make the cuts on the cross-pieces at a very slight angle (a).

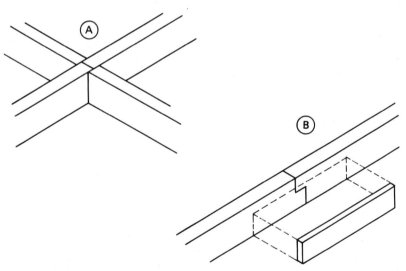

FIG. 19–10   (a) shows ideal way to splice grid pieces when the joint can fall between crosspieces. (b) shows an end-lap to use in other cases or as a means of using up short pieces of stock. Reinforce as shown with ¾ inch stock (scab). Secure the lap joint with a nail driven from the underside.

quality work. It also makes it rougher to do assembly work since you are working with smaller cross sections.

   Regular, dressed 2 × 4's have eased edges; the corners are slightly rounded. It is all right to use them that way, but you'll get a better joint where concrete meets wood if the top edge of the grid piece is squared off. This can be accom-

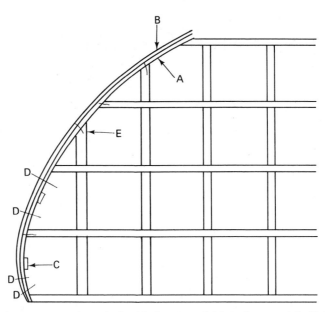

Fig. 19–11 To include curves, work with doubled strips of ½ inch material. Nail the first strip
(a) to the grid pieces whenever possible (e). Provide braces (c) if they are needed.
Add the second ½ inch piece (b) by nailing through it and the first piece and
clinching the nails on the inside of the form (d). Angle-cut the mating ends of the
grid pieces so they conform closely to the curve.

plished by using a table saw to do a rip cut. Some professionals will rip 2 × 8's
in half to get two grid pieces, each with one squared edge.

Check to see if local supply sources have a special material made for use as
concrete gridwork. This will be 2 × 4 lumber in the rough but with one planed
edge. Since this finished edge will be the only one left exposed, it doesn't matter
that the other three surfaces are rough.

INSTALLATION OF GRIDWORK

You can excavate the entire site or start by digging trenches for the perimeter
boards and the main runners. Generally, it is best to set the borders first, at a
height that makes allowance for drainage. This should be a minimum of ⅛ in.
per foot of run. Design the slope outward if the patio abuts the house, toward
a general drainage area if the patio is isolated.

Set the main runners down in one piece if possible. Otherwise, work with
one of the joints or methods shown in the sketches. Stretch out mason's line
for each runner as you go. The line should establish direction and height. It is
assumed, of course, that the perimeter boards have already been set to establish
the drainage slope.

Use plenty of stakes, even more than you feel you need to maintain align-

Fig. 19–12 Stud all concrete-contact areas with 16d galvanized nails. Drive them solidly about midway in the forms; space them about 12-inches apart. They will serve as a bond between concrete and wood.

Fig. 19–13 You can use masking tape before the pour to protect the exposed edge of the grid-pieces. This is more essential if you stain the wood for color before you do the pouring.

ment and hold runner height. Drive the stakes as deeply as you have to for rigidity. When they are solid, drive a couple of nails through the stake and into the runner. Remember that the stakes are permanent and that the excess is cut off at about the center of the runners. This ensures that the top edge of the stakes will be well below the surface of the concrete.

Fig. 19–14 Square top edges on the gridpieces will make a better joint with the concrete. This type of exposed aggregate is done with a sparse seeding of special aggregate. Note the tight joints between gridpieces.

Install crosspieces after all the main runners are in place. Work with a mason's line regardless of whether the crosspieces are staggered or in a continuous line. Cut them to length as you go. If you allow for a few degrees of relief by making the cuts a bit off square, you will get better contact at the top edge where the crosspiece meets the runner; this is the only joint that will be visible after the concrete has been poured.

In-line crosspieces should be end-nailed at one end and angle-nailed at the other. Staggered pieces should be end-nailed at both ends. Use 20d galvanized nails for end nailing, 16d galvanized nails for angle nailing.

Stud all concrete-contact surfaces with 16d galvanized nails spaced about 1 ft apart. Set the nails at about the midpoint of runners and crosspieces and drive them only as much as you have to for them to be secure. These nails will act as binders between concrete and wood and will do much to keep surfaces level.

Run one final check after the gridwork is finished. Check for alignment of grid pieces and for height uniformity. At this point, it won't be difficult to raise any low places or lower high ones, even if it means replacing stakes. Such adjustments can't be made after the concrete has been poured.

BRICK

Brick has been a popular patio-flooring material for centuries. Its durability is fact; its function and appeal have been effectively demonstrated in small and large gardens. Part of the attractiveness of brick is the flexibility of placement. You can work with great precision if you wish formal results or be somewhat casual for rustic effects. So long as structural considerations are met, the project will be long-lived and pleasant to live with.

FIG. 19–15 Different types and sizes of bricks and masonry units may be combined to achieve a particular design. In this case, the curved and straight lines of brick were placed before the paving.

You are not forced to work with common brick. Other types such as pavers or split pavers, Norman and Roman, can be used very effectively. When the project will be subjected to freeze–thaw cycles, it is important to select a brick that is approved for use under such conditions. The trade recommends a unit that is graded SW. Manufactured used brick is acceptable if it meets the specifications, but salvaged units are not recommended unless they have been tested to meet the SW requirements.

PATTERNS

Any of the patterns shown can be accomplished by using units that are 4 × 8 in. in actual or nominal dimensions. The latter must be placed with joints that are filled with mortar or sand. The actual-dimension units are placed "tight" with no joint at all.

You can create different patterns, even combine patterns, but remember that nominal-dimension bricks are designed for use with a mortar joint, actual-dimension units are not. Patterns and overall layouts can be planned to minimize brick cutting; however, the design is so important in the long run that it should not be influenced solely by this one factor.

MORTARED OR MORTARLESS

Whether to use mortar or not is up to you; either way can result in a durable project. If full mortaring is done, the concrete subbase should be at least 3 in. thick, the mortar bed ½ in. thick. When working on a sand bed, be sure that the subgrade is carefully prepared and that the sand thickness is 1 to 2 in. In

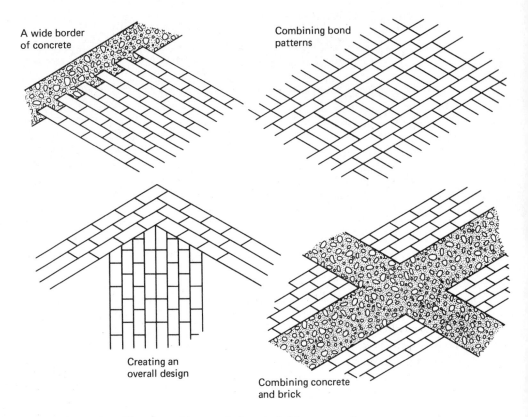

A wide border
of concrete

Combining bond
patterns

Creating an
overall design

Combining concrete
and brick

FIG. 19–16 Examples of how to achieve variations in brick pavement.

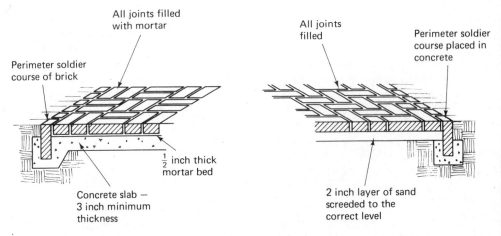

All joints filled
with mortar

All joints
filled

Perimeter soldier
course placed in
concrete

Perimeter soldier
course of brick

$\frac{1}{2}$ inch thick
mortar bed

Concrete slab —
3 inch minimum
thickness

2 inch layer of sand
screeded to the
correct level

FIG. 19–17 Two designs for brick pavement with mortar joints.

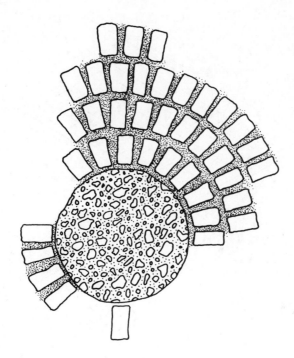

Fig. 19–18 Circles in brick: mortar joints are not uniform in width—best to make a "dry" run for each course before setting the bricks permanently.

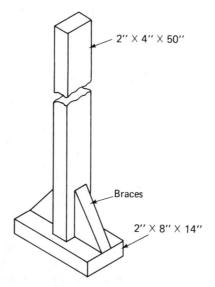

Fig. 19–19 Homemade tamper can be used to help level brick pavement—also usable on sand beds.

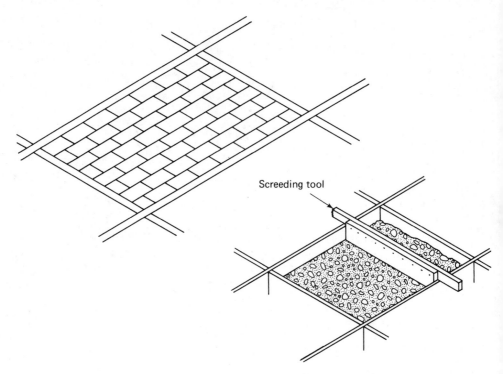

Screeding tool

FIG. 19–20 Brick pavement can be laid in a grid system just like poured concrete. This serves to break up large areas and offers the opportunity to combine bond patterns. To level the bed, make a screeding tool like the one shown above.

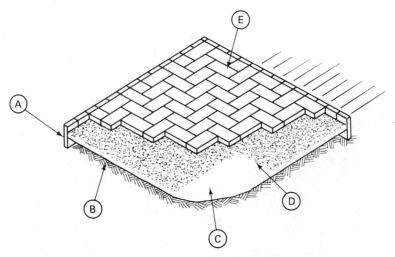

FIG. 19–21 Brick pavement without mortar: (a) edging of bricks or wood; (b) firm earth sub-base; (c) 1 or 2 inch layer of sand screeded to correct level; (d) sheet of plastic film (or 15# felt); (e) fine sand swept over the laid bricks to fill the joints.

addition, consider placing a layer of roofing felt directly over the sand. The felt is not essential but it does provide a smooth bed for brick placement and acts as a barrier to prevent grass or weeds from coming up through the joints.

Mortar joints can be included whether the project is done on concrete or on a sand bed. Chapter 12 gives several methods of doing mortar joints.

DRAINAGE

Good drainage is as necessary with brick paving as it is with concrete. Actually it may be even more critical here because excessive moisture can cause efflorescence or disintegration under freeze–thaw conditions and even promote the growth of molds and fungi. How much attention you pay to below-surface drainage depends on soil conditions. Porous soils that drain quickly are not much of a problem, but clay, adobe-type soils will hold surface water. In all cases, it is wise to slope subgrades and subbases as well as the surface of the project. If the situation is severe, the bricks may be placed on a layer of pea gravel.

A slope of ⅛ in. per running foot will do for small projects; ¼ in. is better for large projects.

EDGINGS

Perimeter bricks, especially when the project is done on sand, can move out of place even under normal traffic conditions. Therefore, a solidly placed edging makes sense. The edging can be composed of wood forms, as you would do for a concrete pour, a concrete footing for the border bricks, a course of soldiers set in concrete, and so on. Whatever you decide to use, do the edging first. This will provide accurate elevation and slope guides when you get to doing the pavement.

PSEUDOBRICKS

Again we make the comment that some products that we place in this category of pseudobrick are real masonry but designed and sized for special installation procedures that make them ideal for particular applications. The example shown here (bricover) may be used indoors or out over almost any clean, solid surface, be it wood, concrete, tile, or metal.

The success of the installation depends a great deal on the condition of the existing surface. It must be solid, smooth, and clean. The presence of grease or oil will prevent good adhesion.

In all cases, be sure to read the instructions supplied with the materials. Use only the mastics, grouts, and sealers, that the manufacturer recommends.

FIG. 19–22 Bricover floor brick may be applied over existing floors indoor or out. Be sure to check manufacturer's recommendations if you plan to use the product in areas of intense heat or direct flame contact.

FIG. 19–23 Bricover is set down in a thin layer of special mastic that is spread with a notched trowel. The mastic sets quickly so only several square feet should be covered each time.

Fɪɢ. 19–24 Grout is mixed in a bucket and then transferred to a special applicator bottle. If you have been careful with the grout-mix, the material will flow freely and smoothly into the joints.

PRECAST UNITS

Precast units are a boon because they can supply a durable, attractive surface with minimum effort. Overall quality is attested to by the fact that precast units are often used for such commercial installations as driveways and municipal sidewalks.

Common precast slabs are square or rectangular in sizes that go up to 36 in. in length. The average thickness runs from 2 to 3 in. As a guide, consider that the minimum thickness is acceptable for patios (or walks) around a residence. The heavier units may be used for such projects as driveways. You are not restricted to squares and rectangles. Other available shapes include rounds, dia-

Fɪɢ. 19–25 Prepare the site for precast units by staking off the area and excavating to the required depth. Depth of the excavation should equal the thickness of the units plus 2- or 3-inches for the sand bed.

FIG. 19–26 Be sure the subgrade is firm and that it is sloped ⅛-inch to ¼-inch per running foot for drainage. Set down temporary, straight boards that you can use as guides for leveling the sand bed.

FIG. 19–27 Shovel the sand into place and then level it by screeding as shown here. Tamp the sand, add more, and relevel it. Be sure that the sand is in a damp condition.

FIG. 19–28 Start from a corner to set the precast units on the sand bed. If you have done the sand bed in good fashion, the slabs will be stable and will provide a lasting pavement.

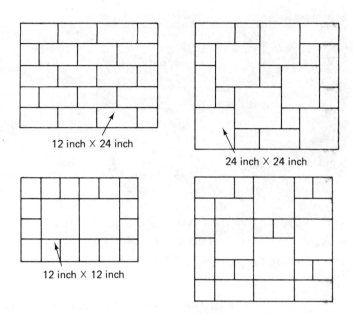

12 inch X 24 inch

24 inch X 24 inch

12 inch X 12 inch

FIG. 19–29 Examples of designs created by using different size, precast patio pavers. The idea can be extended by including various colors.

FIG. 19–30 Precast units (often called patio pavers) are available in many designs. This one is called "Spanish" and is available in half-units as well as different colors.

monds, hexagonals, triangles, even Spanish tile. As they come in different colors and in different textures, you can add interest to any project by thinking beyond the basic pattern bond.

Like brick, the units may be placed with full mortaring over concrete or they may be laid, simply, on a sand bed. The latter is fine for walks and patios,

FIG. 19–31 The procedure for placing them is the same one used for squares or rectangles. In this case, the worker is placing a sheet of plastic over the sand bed.

FIG. 19–32 Heavy roofing felt may be used in place of the plastic sheet, but in either case, the extra material assures a level surface and prevents weeds or grass from coming up through the joints.

but the more substantial concrete subbase makes sense for such heavy-duty installations as driveways.

In any case, there is no doubt that the result will be only as durable and as stable as the subgrade and the subbase work.

Fig. 19–33 This style of patio paver produces a very intriguing design. It's hard to tell by look-
ing, yet all the pieces are the same size and shape—only the color changes.

QUESTIONS

19–1   What are some possible uses that should be considered when designing a patio?
19–2   Name some conditions that should be considered in relation to patio exposure.
19–3   Give some reasons for including gridwork in a concrete patio.
19–4   What is the minimum amount of slope that should be allowed for drainage?
19–5   Why is it a good idea to stud the grids with nails?
19–6   Make a sketch that shows how main runners should be joined when they can't
       be set down in one piece.
19–7   How high should stakes be, and why?
19–8   What is an important factor to consider when choosing a brick size for a patio?
19–9   Make a cross-section sketch that shows brickwork on a concrete subbase and on
       a sand bed. Show the dimensions.
19–10 Why are edgings for brickwork important?
19–11 Make a scaled drawing of a house on a 80 × 150 ft lot. Include a patio and
       indicate materials. Explain reasons for the patio location.

20

Walls

Because well-constructed masonry walls are attractive and will last indefinitely with minimum maintenance, they are much in evidence today. Masonry is being used not only for foundations or building walls but for perimeter fencing, privacy screens, flowerbed borders, dividers, and the like. The popularity of masonry for such projects has increased in proportion to the continuing introduction of new materials, new shapes, new textures, and to the availability, even to the layman, of construction "secrets" that assure professional installation.

Walls of concrete or masonry units no longer have to appear as awesome privacy or security structures. They can be light or heavy in appearance, rustic or formal, solid or pierced. In all cases, they can be designed so that, in themselves or when accented with greenery, they will not be overly predominant.

The honest craftsman can erect just about any wall correctly but certain factors must be considered beyond assembly techniques. Bearing walls and retaining walls, especially, must have built-in safety factors that will assure performance under prevailing conditions. In such cases, the best bet is to make a sketch of what you plan and then to consult a building inspector. Local codes may spell out restrictions, but they will be an excellent guide to a successful project. Be especially careful to abide by rules that relate to the height of free-standing walls. It is almost certain, for example, that acceptable wall heights on the street side will be lower than those you can use on side and back lines.

Fig. 20–1   Brick is durable and attractive. It acquires a special patina and character as it ages. This unusual, undulating wall has been around for a long, long time.

Fig. 20–2   Different types or sizes of masonry units can be used in connecting projects. Good choice of materials results in a pleasant blend, like that achieved in this sunken patio.

Fig. 20–3   Selection of materials, and even design, is unlimited when the wall is small and does not have to stand up under soil or wind pressures. This wall was done with thick slabs of Arizona flagstone.

Fig. 20–4   Simple retaining wall was used to protect the tree when the surrounding grade had to be raised. Pieces salvaged from a razed concrete slab, and natural stones were combined here.

FIG. 20–5   Random cut stones were used here and without a special pattern bond. Careful placement of stones is necessary so that joints will be staggered enough for a good structural bond.

RETAINING WALLS

The size of a retaining wall has everything to do with the pressures that build
up behind it. In essence, the wall must be designed and constructed so that
it can't overturn, slide forward, or rupture at any point. A two- or three-brick
high wall, built to contain soil for a raised flower bed is a fairly uncomplicated

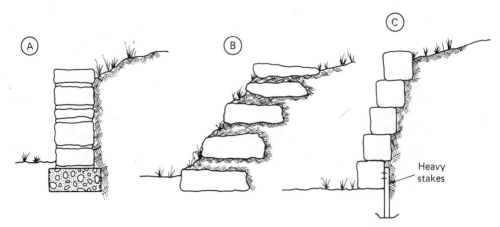

FIG. 20–6   Ideas for retaining walls: (a) cut stone on concrete footing; (b) large field stones
with soil joints—use plants in the joints to hold the soil—count on 3 to 6 inches of
setback per foot of rise; (c) used railroad ties—may be toenailed at ends with heavy
spikes—stagger end joints.

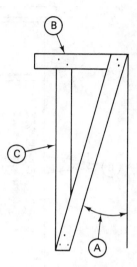

FIG. 20-7 When the wall has setbacks—is sloped—(a) you can make a batter board device (b) to use as a gauge while you work. Use a level against this edge (c).

project. A wall that is put up more to prevent soil erosion on an existing slope than to retain is also pretty simple. On the other hand, with a 3-ft-high wall that must hold back an upward-sloping volume of soil, you enter a sphere of construction in which engineering knowledge is imperative. This is especially true when prevailing conditions include such things as silt, unsettled soil, and pressure problems that must be faced because of water buildup.

A vertical-faced, concrete retaining wall that is 3 ft high and 10 in. thick at the top should have a footing that is 27 in. wide if the soil behind it is level with the top. If the soil slopes upward (has *surcharge*), the width of the footing should be 30 in. The footing for a 10-ft high wall, 10 in. thick at the top and *without* surcharge, should be 63 in. wide; *with* surcharge, the footing width should be 84 in.

These dimensions are subject to change in relation to the design of the wall, which may be vertical-faced, inclined backward, or cantilevered. Cantilevered designs require very special reinforcement procedures. All this points up the wisdom of paying for engineering knowledge when it is needed. Expert masons work from plans that include the information. So should the apprentice and the do-it-yourselfer.

Typical factors to consider when planning a retaining wall are these. Will rain, which may be excessive at times, cause a flow of the soil toward the wall? This can build up tremendous pressure. Are you cutting into an existing bank where the soil has become and will remain stable and where the structure of the soil is not overly affected by moisture? Soil porosity may be such that water buildup behind the wall will be minimal and less significant in terms of wall strength and drainage provisions. An opposite example would be an adobe-like subbase with a layer of loose topsoil. Here, the water would fall vertically

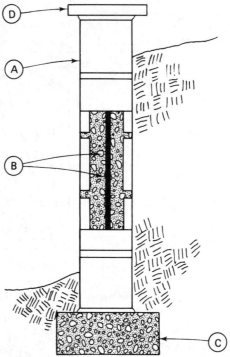

FIG. 20–8   Example of a concrete block retaining wall: (a) 8 inch x 8 inch x 16 inch concrete
blocks; (b) steel reinforcement rods plus cores filled with concrete provide extra
strength—sometimes called for by local codes; (c) concrete footing set below the
frost line; (d) cap.

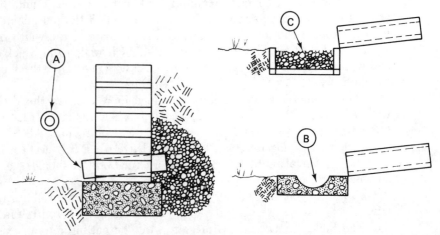

FIG. 20–9   (a) It's possible to get drainage by passing drain tile or pipe through the wall. (b)
To avoid flooding when water is a problem, you can construct a run-off concrete
trough in front of the wall. (c) The trough can be made of redwood, lined with
plastic sheet, and filled with gravel.

through the topsoil but go laterally when it hits the adobe. If the subsurface river is going to hit the wall, you must plan for it.

An easy structural solution is to build at the foot of an existing slope and backfill only as much as you have to to bring the grade to wall level. When the slope has been there for ages, it is not likely that it will suddenly start to bear against the wall; only the backfill will.

A similar situation would be to erect a low wall and then fill behind it to a wall-level grade. Normally, the pressures will be mostly compressive and the wall may be designed accordingly.

In all situations, it pays to include a water drainage system. This may be comprised of drainage tiles placed parallel to the wall and directed toward a general drainage area, or a through-the-wall-duct design that guides water to a runoff area or into a trough.

WALLS OF CONCRETE

Concrete walls require sturdy, well-assembled forms. You will, or should, spend more time with the forming than you do with pouring and finishing. In some situations the footing and the wall are poured as integral units. In other cases, the footing is cast first, keyed to receive the wall pour that follows. In either case the form for the footing may be a trench that is cut directly in the soil or it may be wood.

Common materials for formwork include 1-in. lumber or ⅝- or ¾-in. plywood. Plywood has the advantage of large sheets with minimum joint lines. With 1-in. lumber, it is difficult to avoid the lines that form between the boards no matter how careful you are with the assembly. To avoid this, or simply to get a smoother finish on the concrete, you can line the forms with roofing felt or a sheet of plastic film.

If a unique texture is the goal, you can use as formwork narrow, rough boards or etched plywood. Either of these methods will give the concrete a woodgrain effect. You can also impress the concrete with materials that are nailed to the inside faces of the forms. For example, use ½-round pieces of molding so that the concrete will have equal- or random-spaced, vertical, concave grooves.

Use plenty of vertical and diagonal bracing to support the formwork; this is not the time to be frugal. Make sure that all vertical formwork joints occur over a vertical brace.

If for some reason the wall cannot be completed in one pour, install a keyed, vertical divider to stop the first pour. When the divider is removed, the second pour will tie into the first one.

Use a piece of 2 × 2 or 2 × 4 to settle the concrete as it is being poured. As always, don't overdo the tamping because this may push too much of the large aggregate to the bottom of the forms. It is a good idea to tap the outside

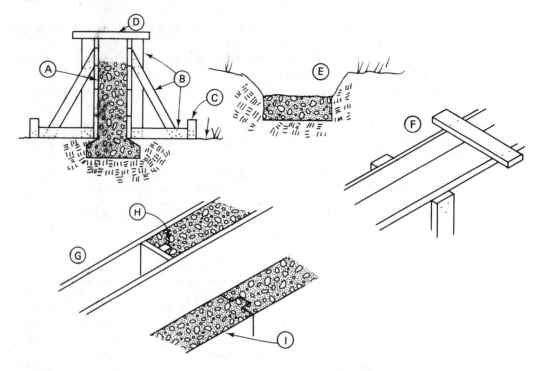

FIG. 20–10 A poured concrete wall requires sturdy forms: Needed are form boards of 1 x material (a), bracing of 2 x material (b), stakes (c), and crossbracing (d). Trenching can be done so wall and footing are integral. Separate footing can be poured in a trench dug in the earth (e) or, when above grade, poured in a simple form (f). When the entire wall can't be done with one pour (g) install a divider (h) with a 2 x 4 key block. The following pour (i) will then be "keyed" to the first one.

FIG. 20–11 Even low retaining walls or planter-box walls may require drainage holes. This is especially true with non-porous soils and situations where top soil has been placed over a clay-like soil.

FIG. 20–12 Low concrete walls that must sustain nothing but themselves do not have to be massive. Even so, adequate footing is required and reinforcement steel is recommended.

FIG. 20–13 Walls can be curved wholly or in part. The finish on this wall was achieved by applying a plaster-like coating of heavy-duty mortar after the forms were removed.

FIG. 20–14 A low wall that is needed to define an area or to retain soil, can do double-duty as a seat. This preassembled wooden cover just slips over the wall—no permanent attachment.

of the forms with a hammer as the pouring progresses. This usually results in a smoother concrete surface.

Generally, it is safe to remove forms after about two days. If the weather is cold, you should at least double the waiting period, if not extend it to about one week. In some situations it is necessary to remove formwork as soon as possible—for example, when you want to expose the surface aggregate. Wait a minimum amount of time and then remove the forms carefully to avoid marring the edges. If setting has progressed enough to make washing and brooming difficult, you can work with a wire brush to loosen the fine aggregate. What you use and how much pressure you apply depends on the condition of the surface.

WALLS OF BRICK

Possible designs for, as well as purposes for, brick walls are unlimited. Free-standing garden walls may be one brick thick and either serpentine or straight. The serpentine wall has a great deal of lateral strength simply because of its shape. For a 4-in. serpentine wall, good engineering practice requires that the following general rules be observed: (1) the radius of the curves should not be more than two times the above-grade height of the wall; and (2) the depth of the curves should not be less than one half the height.

Straight walls get lateral strength through thickness and by the addition of

FIG. 20–15 Brick walls can curve gracefully even though each brick is really a straight line that is tangent to the circle. Of course, the bigger the radius, the easier it is to accomplish a circle.

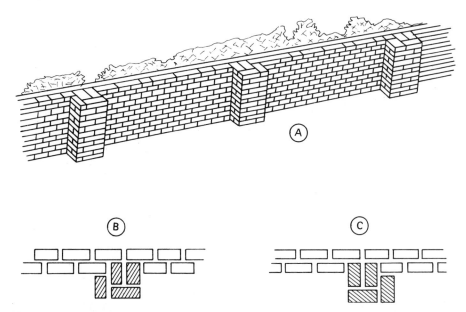

FIG. 20–16 How to do a long brick wall: (a) Integral pairs, called pilasters, are spaced 10 to 12 feet apart on centers; (b) the first course in the pilaster is done this way—note how the design ties units together; (c) this is how to place the second course in the pilaster. Work so with alternate courses until top of wall is reached.

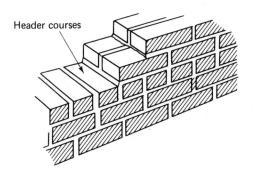

Fɪɢ. 20–17 A standard two-brick wall incorporates header courses.

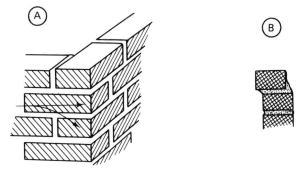

Fɪɢ. 20–18 (a) How to interlock the bricks when turning a corner; (b) with a low, one-brick planter wall, let the top course jut forward a bit to act as a cap.

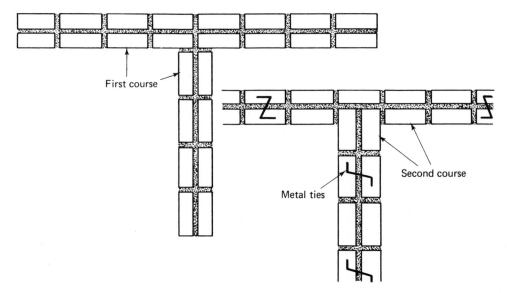

Fɪɢ. 20–19 Basics of an intersecting 8 inch brick wall—with metal ties: place metal ties in mortar joints every 5th or 6th course (stagger).

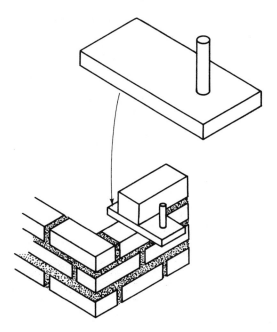

FIG. 20–20 Gate hinge-pins may be set in the wall as you erect it.

special reinforcements. For example, it is recommended that when wind pressure equals 10 pounds per square foot (psf), the above-grade height of the wall not be more than three fourths the wall thickness squared. This assumes that only wind and impact loads are involved and that there is no special bond between the wall and the foundation. At any rate the recommendation for this

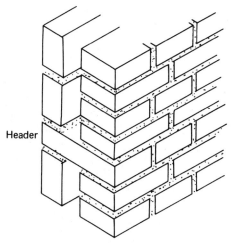

FIG. 20–21 One way to erect a hollow brick wall: span across with a header every fifth or sixth course.

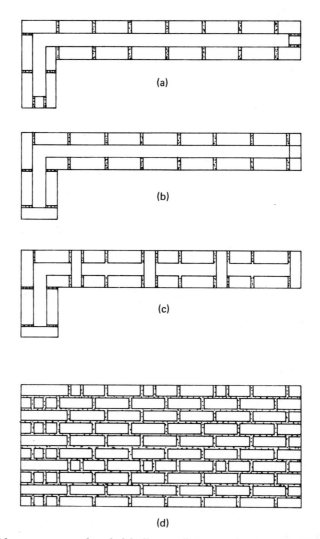

FIG. 20-22 How to build up a masonry bonded hollow wall—example shown is 10 courses high: (a) the first course—the third, fifth, seventh, and ninth are done the same way; (b) the second, sixth, and eighth courses are done this way; (c) the fourth and tenth courses incorporate the bonding units; (d) front view of the complete project.

form of garden wall does point out how much higher you can build with comparatively little increase in wall thickness. The difference, for example, between 4 and 8 in.² is considerable.

The structural requirements of brick-bearing walls are stringently specified by local building codes. In some areas, especially where earthquakes are a hazard, brick is used mostly as a veneer, but in these cases the total height is subject to limitations.

Brick walls may be solid or hollow and in either case may be masonry-

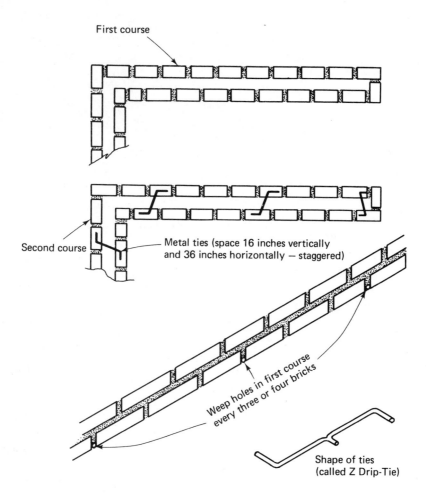

First course

Second course

Metal ties (space 16 inches vertically
and 36 inches horizontally — staggered)

Weep holes in first course
every three or four bricks

Shape of ties
(called Z Drip-Tie)

FIG. 20–23 Basics of a 10 foot cavity wall.

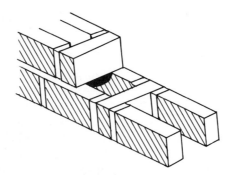

FIG. 20–24 A void wall can be done this way. It's fast and economical but not intended as a
        bearing wall.

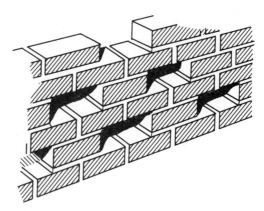

Fig. 20–25 This is how you can do a screen type brick wall.

bonded, reinforced with steel, or both. Hollow walls are often used when drainage problems exist. The usual design allows for water to pass through the facing wall but makes provision to prevent the water from passing through or affecting the backup wall. Trapped water is drained off through weep holes (usually) that are incorporated in the design. Also, a hollow wall may be used to provide space for granular insulation. Of course, a hollow wall may be free-standing if the extra bulk is more esthetically pleasing than a thin wall.

Detached planters may be constructed with walls one brick thick that are set on a perimeter footing or a full slab. When the latter method is used, it is wise to pour the slab on a gravel base and to provide for drain holes. In either case, it is a good idea to waterproof the inside faces of the walls to prevent efflorescence and other types of stains from appearing on visible surfaces. This can be accomplished by lining the inside with a waterproof membrane or by parging.

You can put up pierced walls by working with brick only. Such a project should be regarded mostly as a decorative screen to provide separation between areas without completely blocking out light and air. In some situations a brick, screen wall may be used to support a light overhead structure, such as a carport roof or an overhang for a patio. Numerous patterns are possible because of the different bonds you can use and the different sizes of available materials.

BUILDING A REINFORCED BRICK MASONRY WALL

Preparation for this job begins by having a good mating area on the surface of the slab or footing. It must be clean, and the coarse aggregate should be exposed. The vertical steel is placed accurately when the footing or slab is poured. The horizontal steel is not tied but floated in the grout as the job progresses. When you place the mortar bed for the first course of brick, be sure that you confine it to brick areas only. The bond in the center of the wall should be between the foundation and the grout.

MORTAR FOR REINFORCED BRICK MASONRY WALL

Portland Cement	Parts by Volume Hydrated Lime	Sand
1	¼ to ½	2¼ to 3 times the sum of cement and lime

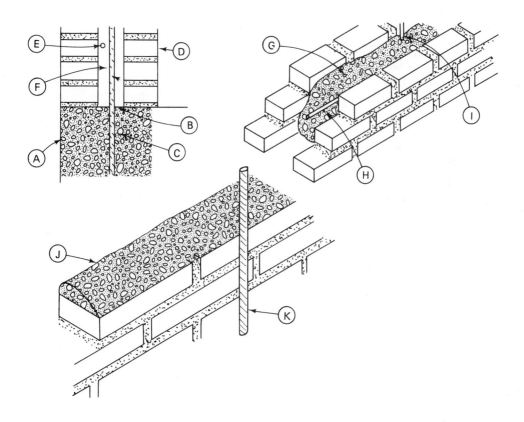

Fig. 20–26 The basics of a reinforced brick masonry wall: (a) concrete footing, or slab; (b) mating area clean and with aggregate exposed; (c) vertical reinforcement rod centered in cavity and spaced about 24 inches; (d) bricks; (e) horizontal reinforcement rod in third course and every fifth or sixth course thereafter; (f) cavity to be filled with grout (see chart for correct grout mix); (g) pour grout every third course—keep ½ inch below level of brick—puddle with stick to be sure grout fills cavity completely; (h) place horizontal reinforcement rods in grout; (i) be sure that vertical rods remain centered; (j) use a beveled joint for mortar—do not furrow—be sure that excess mortar does not fall into cavity; (k) vertical rod.

CAVITY GROUT FOR REINFORCED BRICK MASONRY WALL

| | Parts by Volume | | | Aggregates | |
Type	Portland Cement	Hydrated Lime	Sand	Coarse (⅜ in max.)	Use
Fine	1	0–1/10	2¼–3 times the sum of cement and lime		For cavities not more than 2 in. wide
Coarse	1	0–1/10	2¼–3 times the sum of cement and lime	1–2 times the sum of cement and lime	For cavities more than 2 in. wide

Bed mortar for the remaining courses of brick should not be furrowed. Instead, as shown in the sketch, place it in beveled fashion. The idea, throughout, is to keep excess mortar from falling into the cavity and mixing with the grout. Be very careful to get full joints and to keep all steel centered.

Don't wait longer than three courses before filling the void with grout but do allow at least 15 minutes between successive grout pours at any point in the wall. When grouting is done too quickly, you can build up sufficient pressure to bow the wall or, as the trade says, cause a "blowout." In fact, experienced masons will not work more closely together than 10 or 15 ft so as to avoid concentrated pressures.

Puddle the grout immediately after it is poured. Don't make puddle sticks from lumber that is more than ¾ in. thick and 1 in. wide. Sticks that are larger, and puddling that is not done immediately, can cause structural problems, even contributing to blowout.

WALLS OF BLOCK

Concrete blocks will produce sturdy and economical walls and the big 8- × 8- ×

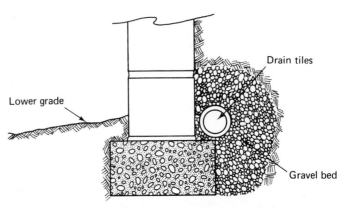

Fig. 20–27 Drainage of walls. Drain tiles are used to carry away water that seeps to bottom of wall. The tiles are placed in a bed of gravel.

FIG. 20–28 Concrete blocks or similar units can serve very nicely as retaining walls. In this application, the blocks were reinforced vertically with steel and concrete-filled cores.

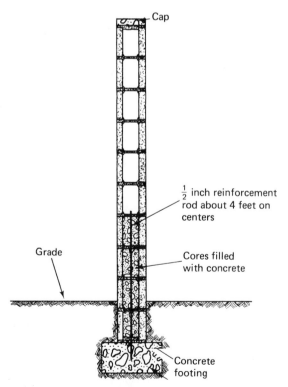

FIG. 20–29 A masonry unit screen or privacy wall for the garden. Do the block work following recommended practices—keep above-grade wall height to 6 foot maximum—reinforce the lower 2 feet above grade as shown.

16-in. units contribute to fast construction. Their appearance can be simple or ornate, depending on the types of masonry units that you choose and how much work you wish to put into doing a special pattern bond. Running bonds, stack bonds, ashlar patterns, and the like—all can be effective when combined with similar or contrasting units.

The technique for laying block doesn't differ from that for laying brick, but reinforcement procedures, when required, may be special in relation to what the wall must do. As with brick, structural specifications vary in different areas of the country. Again, it makes sense to appear before the building inspector, sketch in hand, before starting the job.

WALL·CAPS

Walls are topped for appearance and for protection of the masonry. The job can be cast in place using a heavy-duty mortar or a concrete mix with minimum-sized, large aggregate, or by forming the topping freehand. Other methods

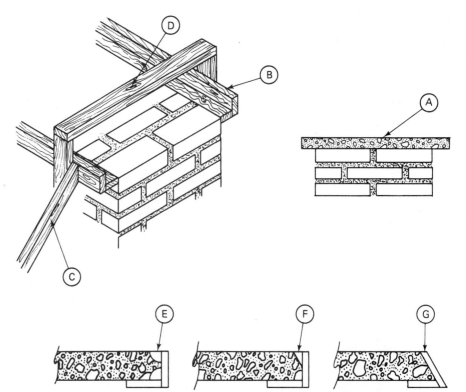

FIG. 20–30 You can cast an integral cap on brick or block walls (a). Make the side forms using 1 x material (b) and use side braces to the ground (c) and cross braces above (d). Examples of fancy edges you can get are shown in (e) with quarter-round molding and (f) with triangular pieces. Or you can create sloping edges, as in (g).

FIG. 20–31 The kind of cap you use depends on the appearance you want and the amount of time you wish to spend doing it. This masonry unit wall was topped simply with precast slabs set in a bed of mortar.

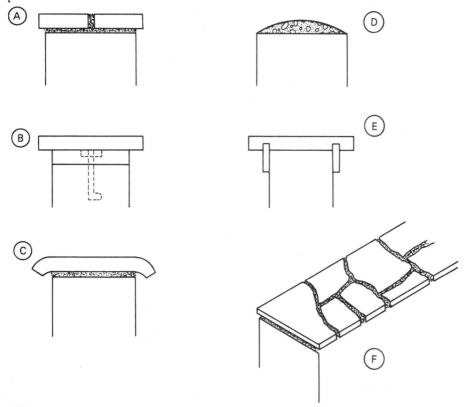

FIG. 20–32 Other ways to cap walls: (a) bricks; (b) nail a 2 x wood seat to a board that is attached to the wall with an anchor bolt; (c) commercial caping; (d) round off by hand using stiff concrete mix; (e) preassembled wood seat; (f) flagstone—or you can use tiles.

include using similar or contrasting materials set in a bed of mortar—for example, brick, thin precast slabs, flagstone, or readymade copings that can be concrete or ceramic. Wood makes an interesting topping when the wall is low and can serve as a bench. A wood topping can be anchored to the wall or preassembled and set in place without fastening it to the masonry.

In any case, and especially when the wall is thick or will be used as a bench, organize the top to incorporate a degree of slope that will prevent water from accumulating.

STONE WALLS

Most stone walls are identified either as *rubble* or as *ashlar*. The terms describe the kind of stone that is used and the manner in which it is placed. Rubble is composed of uncut stones that you may buy or pick up in a field. The pieces, as found, are fitted into the structure as uniformly as possible but keyed in such a way that they form a structural bond.

Ashlar is the term for stones that are cut to fit in a particular pattern bond

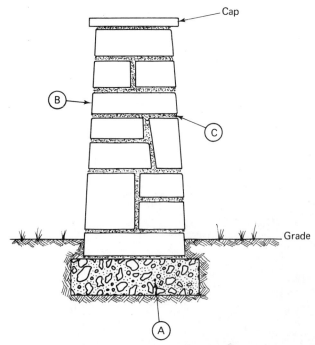

FIG. 20–33 Example of a cut stone wall: (a) poured concrete footing should be about 12 inches thick and wider than the wall by about 8 inches—set below the frost line; (b) header stones, as in brick work, act as ties; (c) good mix for the mortar joints consists of one part cement, one-half part of fire clay, and three parts of sand.

that is also structurally sound. Ashlar is easier to work with than rubble, but its cost can be quite high.

Rubble is much cheaper but can be tricky to use. If you try this technique, be sure to place most of the larger stones in the base courses, fitting the stones together as closely as possible; don't rely on mortar to fill large spaces. Select appropriate stones to be used as headers.

Stagger the joints as you would do with brickwork but don't stick to a rigid pattern. Joints cannot be uniform but a skilled stoneworker can assemble uncut stones so that they appear to have been made for the job. You can do the same if you work slowly and are patient.

This type of work is still done "dry," that is, without mortar. Dry, fieldstone fences, erected many, many years ago by farmers who were clearing land, are still in good condition.

QUESTIONS _____

20–1   Make a drawing of a concrete retaining wall that is 3 ft high and 10 in. thick at the top. Show the footing dimensions when the soil behind the wall is level *and* when there is surcharge.

20–2   Make two sketches that show how a water drainage system can be designed in a wall.

20–3   What can you do to get a smooth finish on a cast wall when the forming is done with 1 in. lumber?

20–4   Make a drawing that shows a keyed, vertical divider in a wall form. When is such a divider used?

20–5   What are the general rules to follow when doing a brick, serpentine wall?

20–6   What are the recommended specifications for a brick wall when the wind pressure equals 10-PSF?

20–7   What factors in relation to the above question are involved?

20–8   Make a cross-section sketch that shows a small brick wall that is one brick thick and three bricks high. Show the footing and indicate the dimensions.

20–9   What is a brick *screen-wall*?

20–10   What ingredients do you use when mixing the mortar for a reinforced brick masonry wall?

20–11   Name the ingredients for a *fine* cavity grout.

20–12   When should you use a fine cavity grout?

20–13   When should you use a coarse cavity grout?

20–14   What extra ingredient is used in the coarse grout? What amount?

20–15   Why is it suggested that the bed mortar be *beveled* instead of *furrowed*?

20–16   Make a cross-section sketch that shows a 6 ft high concrete block wall and its footing.

20–17   Name some materials (or methods) that can be used to top a wall.

21

Ups and Downs of Steps

Steps are always included in a landscaping plan because they are necessary. Unfortunately, therefore, they are often seen as a necessary evil, an approach that is apt to result in a very dull solution. While it is true that the basic function of steps is to get traffic from one level to another conveniently and safely, they can play an even more important role both functionally and visually.

Long steps can expand a small veranda, porch, or raised terrace. Conversely, narrow steps, especially when placed in a severe line, do nothing to supplement a feeling of spaciousness. It is wrong to feel that such thoughts apply only when the lot is large. The whole point is that visual impact is as effective, indeed is even more important, on small lots. There is no doubt that the designer has to live with space restrictions, but he can view them as a challenge to greater creativity as well as craftsmanship.

Steps can even create a mood. Long and gentle, they suggest a leisurely approach; narrow and steep, they hurry you—your mind is more on getting up safely than on enjoying what may be nearby. The former steps form part of the scene. The latter are a means of getting from one scene to another.

Any of the materials we've discussed, as well as others such as salvaged railroad ties and heavy beams, can be used in step construction. Brick and concrete, alone or in combination, are very popular. Full-sized concrete blocks are not seen too often, possibly because of the set dimensions they impose, but they can be

FIG. 21–1 Poured concrete steps can be finished in a variety of ways. This is a kind of halfway exposed aggregate achieved by using a special aggregate in the mix itself, and by washing after forms are removed.

FIG. 21–2 Precast concrete slabs are often used as treads. This commercial concept is almost a "kit" project that includes the understructure and the slabs. It's also available for home installations.

Fig. 21-3 Slab-treads may also be used over masonry unit subbases. Such slabs may be purchased ready to use, or they may be cast on-site, dimensioned exactly for the project on hand.

used effectively alone or with other masonry. Results can be quite formal or very rustic. Natural materials are often included in the design. Stones or boulders already on the site (or trucked in from elsewhere) may be used like raised sideboards to contribute a natural feeling.

It is obvious, therefore, that the many materials you can choose, combined

Fig. 21-4 It might make sense to consider a ramp instead of steps. Justification may be a very slight change in elevations or an easy-to-use path for wheelbarrows, lawn mowers and the like.

with flexibility in design, makes it possible to construct steps that are safe and practical while suiting any landscaping scheme.

RISER—TREAD RELATIONSHIP

The height of the riser and the depth of the tread are very important. A *correct* relationship should be established between the depth of the tread and the existing

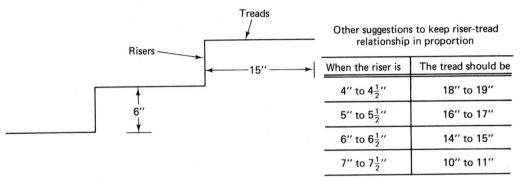

Other suggestions to keep riser-tread relationship in proportion	
When the riser is	The tread should be
4″ to 4½″	18″ to 19″
5″ to 5½″	16″ to 17″
6″ to 6½″	14″ to 15″
7″ to 7½″	10″ to 11″

Fig. 21–5   This is a good relationship between tread and riser for outdoor steps.

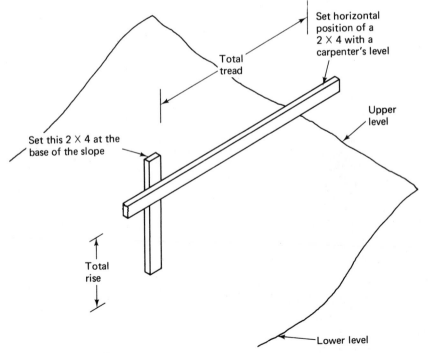

Fig. 21–6   Checking the existing slope.

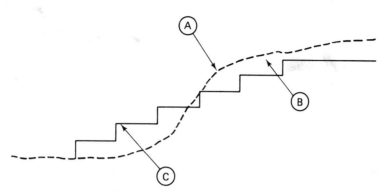

Fig. 21–7 If building the steps to fit the existing slope (a) would result in a bad riser-tread relationship, you can make changes by cutting into the slope (b) or by building out on the lower level (c).

slope or total rise if possible, but *established*, even if the dimensions do not conform with existing conditions. This would be very restrictive if there were but a single riser–tread relationship acceptable. But since there are many, it is usually possible to work out a solution that is both safe and attractive and that requires minimum extra work in terms of grading.

The problem is particularly acute when you wish to step up an existing slope. Letting the slope angle dictate how steep or how shallow the steps must be is not the procedure to follow. If the steps are too steep, you may have to include handrails for safety. If they are too shallow, a gentle ramp may be preferable. There is a lesser problem with an abrupt change in elevation—for example, steps required from grade to the first floor of a house. In this case, the total rise is fixed, but you can choose the total tread run, which you can vary according

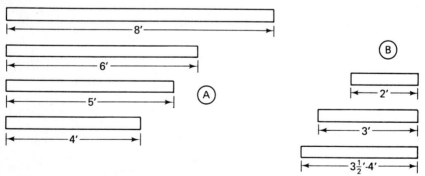

Fig. 21–8 How wide should steps be? (a) Landscaping steps should be a minimum of 4 feet wide if they are to accommodate two people walking side by side. The wider you make them the more important they become in terms of visual impact. (b) Utility steps, necessary but unimportant in the landscaping, must still be a minimum of 2 feet wide to accommodate one person. The 3½ to 4 feet width is very generous but the 3 feet width makes sense if the steps are to be used for carrying garden equipment.

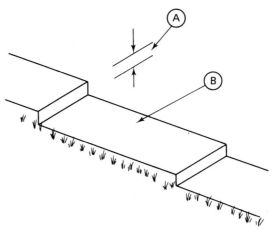

FIG. 21–9 Guide dimensions for a stepped ramped walk on a long slope: (a) keep riser **height** to about 6 inches; (b) do not slope the treads to less than ⅛ inch per foot but **try** to stay under ¼ inch per foot.

to the limits of safety and the unlimited ideas that you may have for an attractive design.

Generally speaking, wide treads go with a short rise. If you step high, you tend to bring your foot down vertically. Because you use a more natural walking stride when the rise is low, you need more tread depth. There will be situations where you must do the best you can in the available space, but do

FIG. 21–10 Steps that are installed for utility purposes only can be minimal in concept but should be as safe to use as any others. Here, permanent wood sideboards were part of the original formwork.

remember that steps do not always have to go in a straight line. For example, they can turn or pause at some point at a landing.

The best way to proceed is to first determine the height between the two levels (total rise) and then the horizontal distance (total tread) that you have to work with. Work on paper with these figures, doing a very simple scaled drawing. When you arrive at a satisfactory tread–riser relationship, you can start to think of design.

DESIGN

After you have given adequate thought to safety factors, think of the steps as a design element that can do double duty. For example, steps may be constructed

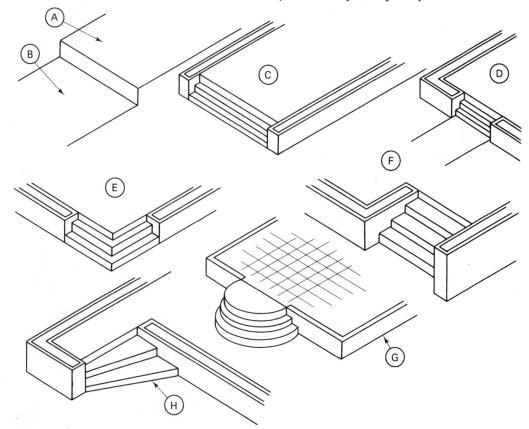

FIG. 21–11 A few ideas for step designs: (a) upper level; (b) lower level; (c) total-width step—side walls can be solid or cavity-type walls for use as planters; (d) minimize step width by wrapping walls around corners—may also be done by cutting into an existing slope; (e) the steps may turn a corner; (f) traffic direction may dictate placement at one side; (g) circular steps permit entrance from any direction; (h) diagonal steps can add interest but tread width at the narrow end must be wide enough to be safe.

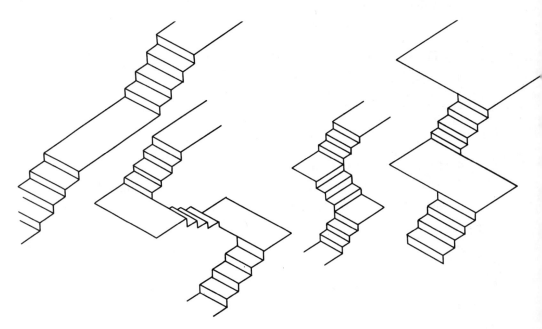

FIG. 21–12 When total rise is substantial, you can design for more safety and better appearance by including landings, ramps, or terraces in the planning.

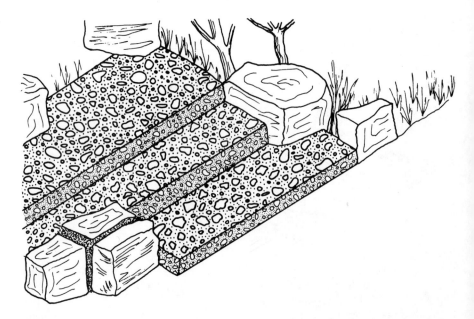

FIG. 21–13 Outdoor steps should be functional but can be designed to enhance the landscaping by using existing situations or by "faking" them. Here, for example, are poured concrete and exposed aggregate steps situated among large stones, shrubs, trees, etc.

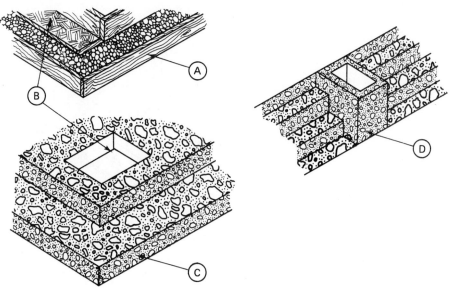

F‌IG. 21–14 Imaginative touches will do much to beautify outdoor step designs—try to keep in tune with the outdoor feeling: (a) permanent 2 x redwood forms with very coarse exposed aggregate; (b) open areas for plantings; (c) poured, exposed aggregate concrete; (d) "built-in" concrete planter to divide very wide steps—may also be used to direct traffic.

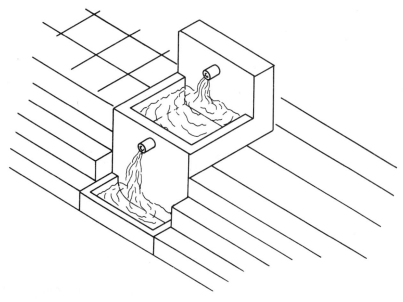

F‌IG. 21–15 Unique landscaping results from imaginative design. It means more work, but the results are worth it. "Built-in" fountain utilizes recirculating pump to return water to upper pool. (See chapter on concrete crafting for information on fountains.)

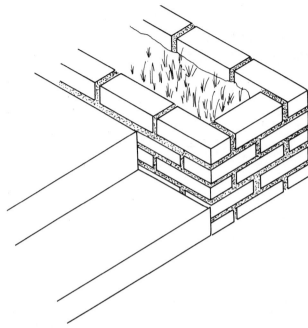

FIG. 21–16 Concrete steps flanked by brick planter boxes. Do the brick planters first and use them as "side forms" for the concrete.

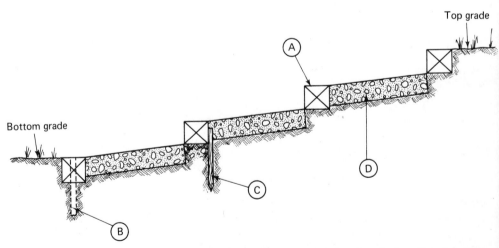

FIG. 21–17 Follow a slope. Risers (a) may be heavy beams or used railroad ties. Set the riser beams by driving pipe into the ground through pre-drilled holes (b) or by using heavy stakes (c). Excavate and pour the concrete treads *after* the riser beams are installed (d). NOTE: The design works best on a gentle slope.

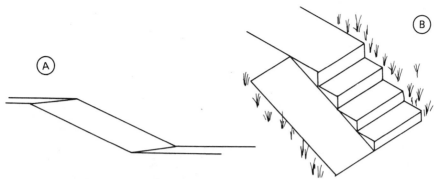

FIG. 21–18 On shallow slopes (a) a ramp may be more functional than steps, especially if traffic will include mowers, wheelbarrows, etc. You can also combine a ramp with steps (b)—the ramp for wheeled traffic, the steps for walking. NOTE: good ramp slope is 1 to 2 feet for every 10 feet of rise.

to serve also as a retaining wall; they can separate areas or establish a visual as well as a physical joining; they can be prominent or concealed with walls, screens, or greenery; they can limit traffic direction or invite use from several directions; they can make narrow places look wider or dispersed areas more cozy; they can be barren climbers or combined with midpoint or border planters to look primarily decorative.

Base your design on elements characteristic of the outdoors. You are not limited to the walls, wells, and floor-to-floor rise typical of indoor steps. Steep steps may be necessary indoors because of space limitations; outdoors, however, there is room for deep treads, and they can provide extra seating. They will definitely be safer under any conditions.

The ideas shown here (in the sketches) may be used as is, or parts of them may be incorporated into something original. Most important is the preplanning stage; it is at this point that you will delineate clearly the function of the project and add touches that will result in something extra. Preplanning can also decrease overall time and effort. For example, when building concrete steps, it takes but a bit more formwork to provide for a built-in planter or even a pool. When planning your project, consider also the installation of underground electrical cable if you wish to have the safety or atmosphere of lighting as part of the project.

STEPS OF CONCRETE

Steps of poured concrete are extremely heavy, so footings for them should be prepared in line with building-code regulations as they affect concrete generally. General recommendations state a footing that is between 18 and 24 in. deep in firm, undisturbed soil, or at least 6 in. deeper than the prevailing frost line. On

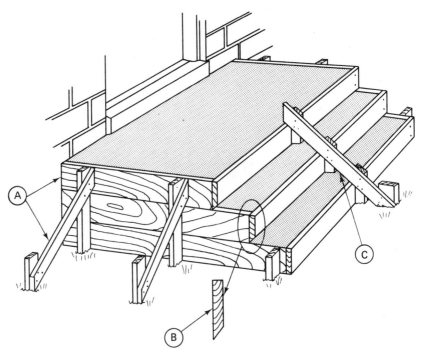

FIG. 21–19 Basic forming for poured concrete steps: (a) okay to work with 1 x material for both forms and bracing; (b) if you bevel the lower edge of the riser boards, it will be easier to do finishing work on the treads; (c) center supports are needed on wide steps—remove just before finishing—fill holes by hand with small trowel.

FIG. 21–20 Don't be frugal with riser-bracing. The longer the steps are, the more of this you need. Note that the sidewall forms are studded panels and that they are well braced.

FIG. 21–21 Steps of this nature are usually cast after the building is up, but the footings may
have been poured as an integral part of the foundation walls. Use special tie-in procedures when step foundation is poured separately.

new work, such footings may be poured along with foundation work. When
the steps are planned as an addition, steel reinforcement should be used to attach the new concrete to existing walls. This can be accomplished with foundation steel that is bent outward to receive the new pour or by drilling into existing walls and using anchor bolts. In addition to metal anchors, some contractors
dig 6- to 8-in.-diameter holes under the front tread location and fill them with
concrete to act as support piers. The whole idea is to keep the new work from
sinking or moving away from the house. The construction of freestanding units
deserves similar care, but the size and depth of footings (if any) must be decided on in relation to local conditions, soil stability, and project design.

Formwork should be only as much as you need to support the concrete without bulging. Remember that the easy stripping of forms without damage to the
concrete is an asset, so work with good material that is straight and free of holes

FIG. 21–22 All risers should be level laterally but they should be positioned to provide for the
tread slope. This should be from back to front and should equal about ¼-inch per
foot.

FIG. 21–23 After the concrete is poured and has set a suitable amount of time, the tread surfaces and landings may be float-finished. An edging tool is used to define the line between tread-front and riser board.

that could cause the concrete to lock to the wood or cause visible imperfections.

How fast you remove the forms depends on the kind of finishing you choose to do. Some masons will remove the riser boards as soon as possible so that they can float-finish riser surfaces, treads, and landings quickly. Others will finish treads and landings as soon as possible but leave all formwork in place for sev-

FIG. 21–24 Be very careful when you remove the riser boards. It will be almost impossible to do this without some damage to the riser faces. An oil coating or a form release will minimize it.

FIG. 21–25 The sooner you remove the riser boards, the easier it will be to float-finish the riser faces with minimum adding of patch mortar. Be sure that the concrete has set enough to support its own weight.

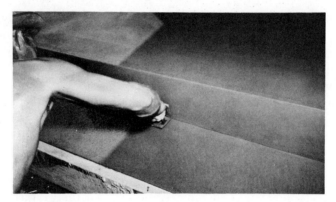

FIG. 21–26 Finish the joints between risers and treads, and the forward edge of the tread, so they are nicely rounded. Use such tools (see next photo) with just enough pressure to sink any surface aggregate.

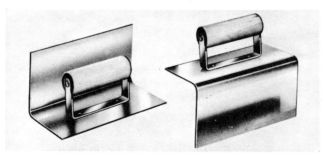

FIG. 21–27 These are among the tools that are used to finish concrete steps. The one on the left does the joint between the riser and the tread—the other is used to round off the front edge of the tread.

FIG. 21–28 The face of the riser may look like this after you remove the riser board. If working with the float only doesn't produce the finish you want, apply a mortar mix to fill voids. Work with the float until the added mortar blends in.

eral days. If necessary, patching can be done after stripping. Sometimes, the riser faces and the sidewalls are coated with a mortar plaster.

Wash finishes, broom finishes, exposed aggregate, and the like, all require that forms be removed as quickly as possible. This can be done anytime during several hours after the pour—it depends on the weather and how fast the concrete sets. Just be sure that the concrete has hardened enough to support its own weight.

To save on the amount of concrete, use rubble inside the formwork—stones, bricks, or broken concrete. Be sure that the fill is compacted and, to be safe, that open spaces are filled with sand. The space between fill material and formwork should never be less than 4 in.

Poured concrete can also serve as a base for other materials. The steps are cast as just described but only as a base for brick, flagstone, precast slabs, and so on. If you do this, be very careful to include the thickness of the veneer material you use when planning the formwork for the concrete base.

Nonslip finishes are a must on steps. Bear this in mind when you choose a material or a particular finish for concrete. Brooming, swirl-floating, and washing are all acceptable. Commercial projects often include special abrasive strips and nosings that are embedded in tread surfaces. Such items are available for home projects and are wise to consider when the situation forces a pitch that is steeper than you'd like it to be. Also consider handrails, either for looks or for safety.

Nonslip finishes themselves, or foot sockets to receive them, are usually easy to install as the masonry work is being done.

STEPS OF BRICK

There are three popular methods of setting down brick steps: (1) a concrete subbase in the form of steps is poured, as just described, and the brick is used as a veneer; (2) a concrete slab is poured as a base for the entire area of the steps; (3) a perimeter footing is poured for what will be the sidewalls of the project.

With methods (2) and (3), one begins by laying down perimeter brick courses and then filling in between. When the job is on a slab, you can fill solidly with brick or with concrete and rubble. With a perimeter footing, the concrete and rubble technique is usually used. Don't build up sidewall courses too high before doing the filling or you may build up enough pressure to bow the freshly placed bricks. To be safe, fill about every second course.

The strength of the project depends a great deal on the subbase work. So

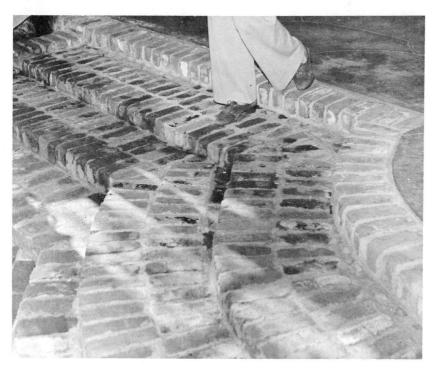

FIG. 21–29 Brick steps require good subbases if they are to be safe and durable. Lines may be straight or curved or, as here, you can combine the two. Design for good tread-riser relationship as well as appearance.

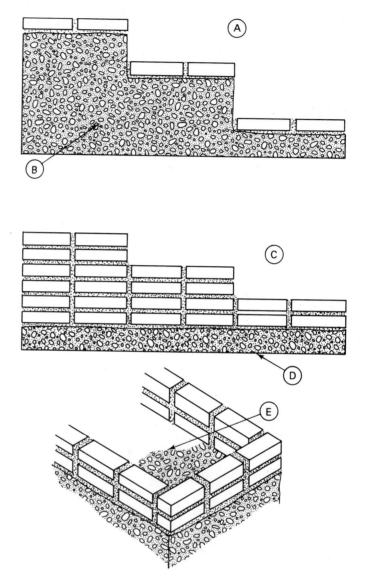

FIG. 21–30 Brick steps can be set down over a poured concrete base (a). You can fill out the concrete pour with rubble (b). Brick steps may also be built up (c) on a concrete footing (d). Fill in with concrete as you build up the brick (e).

do it well. The tread surface should have a pitch forward of about ¼ in. per ft, but this is always provided for topside; the undersurface of a concrete base is stepped off horizontally.

Never use anything but full bricks at the front edge of the treads, and if you extend them as a nosing over the riser, limit the projection to about ½ in. It is best to work with brick that is graded SW and with type M mortar. The

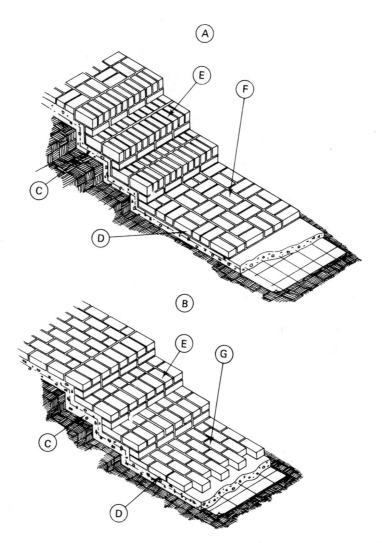

FIG. 21–31 Two designs for brick steps: (a) tread bricks laid on edge; (b) tread bricks laid flat; (c) 4 inch minimum thickness concrete base—use wire mesh as reinforcement when subbase is not stable; (d) tread bricks and those in upper and lower pavement (or landings) are laid on ½ inch thick mortar bed; (e) treads should be at least 12 inches—always use full-length bricks at the front; (f) pavements (landings) done in basketweave pattern; (g) pavement (landings) done in running bond pattern. NOTE: slope treads about one-quarter inch but keep concrete base level.

latter consists of 1 part of portland cement, ¼ part of hydrated lime, and 3 parts of sand. Be sure that all bed and head joints are full. Use a steel tool to compress joints as soon as the mortar has set enough to take a thumbprint. A slightly concave joint will do a good job and will be easy to keep clean.

As we will show later, brick is a good material to use in combination with other items.

STEPS OF BLOCK

Concrete blocks are okay to use as steps if you provide adequate subgrades and subbases. The choice of block type is important since exposed webs are not too pretty. Corner blocks, for example, are good to use since they can be positioned to provide a solid riser face. The height of the riser can be controlled by how you set the blocks. For example, the base for the first tread can be established 3 in. below grade to get a 5-in. riser height. This can be continued with each step that you build up.

Hollow-core blocks can be used imaginatively. Midpoint or border units can be placed with cores exposed topside. The openings make pretty good containers for small plants even if the units are sitting on a concrete slab.

Blocks are often used as sidewalls for concrete or brick steps or as a base material for brick, flagstones, precast slabs, and so on. When used as such, blocks can follow a stepped concrete subbase, or they can be built up on a concrete slab. When used as sidewalls for other materials, the blocks can be erected on perimeter footings. To provide strength when the blocks are used in this way, you can incorporate vertical steel in the footing and fill the open cores with concrete or mortar.

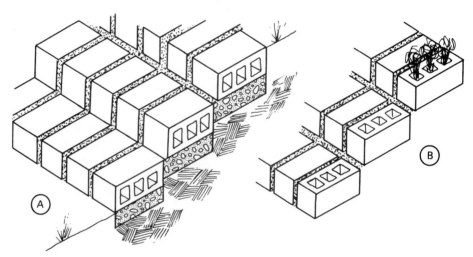

Fig. 21–32 (a) Steps made of 8 inch x 8 inch x 16 inch concrete blocks; (b) set some of them so the cores are exposed. Fill cores with soil and small plants.

COMBINING MATERIALS

Concrete, brick, and block, used in traditional fashion, are all good materials for outdoor steps, but you can create unique effects by combining them with other materials. It is a pleasant challenge to blend structural elements into a

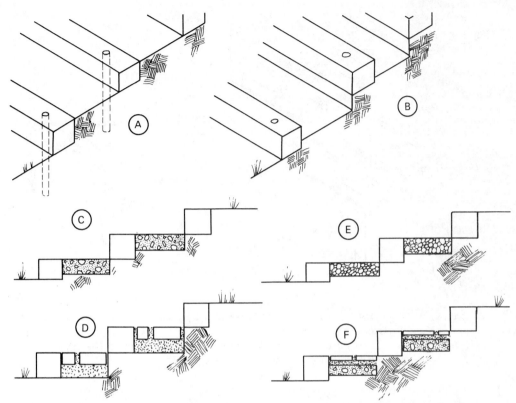

FIG. 21–33 Ideas for steps when railroad ties or heavy beams are used as risers. Set beams as shown (a). Be sure supporting ground is firm. Drive pipe through pre-drilled holes to secure beams. Excavate between beams (b) for tread material, which can be poured concrete (c), bricks placed on tamped sand (d), loose gravel (e), or flagstone set in mortar over a concrete bed (f).

landscape so that the project is compatible with nature itself. Often, and especially on existing slopes, the project can have a built-in look that minimizes the intrusion of something man-made.

Masonry materials in themselves have a kinship with natural things and are not difficult to blend with trees, shrubs, flowers and grass. The same is true of such materials as railroad ties, salvaged beams, wood blocks and rounds, logs, and field stone.

In many cases, you can use heavy wood as formwork for the stringers and risers and then another material for the treads. Depending on personal preference and existing conditions, the tread can be anything from conventional concrete to grass.

Of course, you must never neglect the structural aspects. "Pretty but weak" is not a good combination for our purposes. Be sure of your subgrade and sub-base work and don't overlook technical considerations. The riser–tread relation-

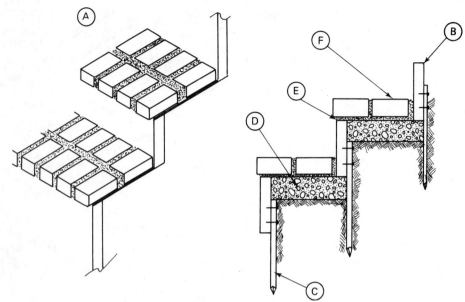

FIG. 21–34 (a) Wooden riser, brick tread combination; (b) risers—2 x 6 redwood; (c) stakes —to hold risers firm; (d) 4 inch thick concrete bed; (e) ½ inch thick mortar; (f) bricks laid flat or on edge. NOTE: Do project by first installing the risers and excavating for the concrete bed—be sure subbase is stable.

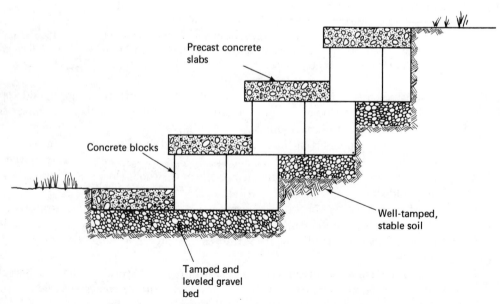

Precast concrete slabs

Concrete blocks

Well-tamped, stable soil

Tamped and leveled gravel bed

FIG. 21–35 Mortarless slab steps. NOTE: Start construction from bottom step and work up.

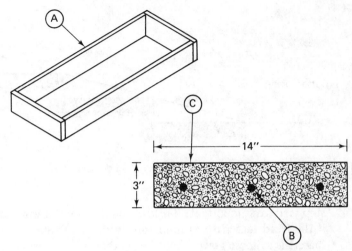

FIG. 21–36 Casting the slabs for the mortarless slab steps. (a) Make a bottomless box for the pour—set the box on a flat surface; (b) use ⅜ inch reinforcement rods—in three places; (c) broom the surface or wash lightly to expose the aggregate.

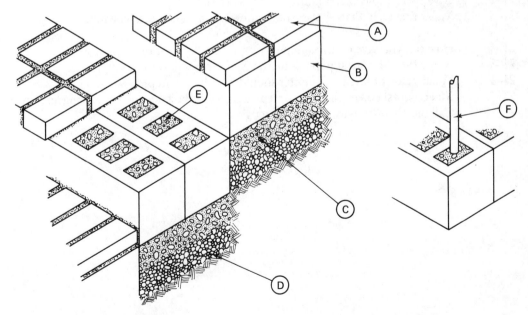

FIG. 21–37 Brick on block: (a) Bricks on mortar bed—to reduce riser height place bricks on edge; (b) concrete blocks set in mortar 4 inch concrete (c)—use reinforcement if soil is unstable; (d) 4 inch gravel subbase; (e) fill cores with concrete or mortar; (f) posts for handrails may be inserted in the cores. Bricks must be cut to fit around the post.

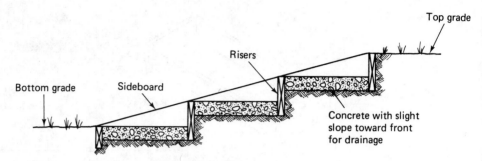

Fig. 21–38 Steps on a slope. NOTE: Sideboards and risers may be preassembled and set into excavation.

ship is always important: Treads should slope outward about ¼ in. per ft even if the tread material is a miniature lawn. Remember that treads of such porous materials as loose gravel should have drainage provisions built in underneath.

QUESTIONS

21–1   Give some examples of good riser-tread relationships.
21–2   Give two reasons why a ramp may be wiser than steps.
21–3   Make a sketch that shows the formwork for a simple, concrete-steps project. Show all bracing.
21–4   How can you save on the amount of concrete required to do solid steps?
21–5   Name the three most popular methods of setting down brick steps.
21–6   What grade of brick and mortar should be used for steps?
21–7   Make two sketches that show steps that were designed by combining materials. Call out the materials and the dimensions.

22

Attachments

There are many items that can or must be attached to masonry walls and floors. These can range from lightweight picture frames and decorative wall plaques, to wrought iron rails and heavy wooden gates, to wall-hung furniture and air conditioning units. The holding material may be hollow or solid, of soft or of extremely hard density. Often, the need for the attachment device can be anticipated. Other times it must be added to an existing structure. Regardless of the circumstances, it is safe to say that you will be able to find a fastener or a particular device suited to the job on hand.

All attachment devices may be listed in one of two main categories—those that you attach during construction, and those that you add after the job is completed. The first category includes anchor bolts, post brackets, steel plates, and inserts for concrete that provide threaded sleeves for machine bolts and such. The second category includes common picture-frame hooks, toggle bolts, "Molly" bolts, expansion sleeves, and special anchors.

Generally, inserts are placed to provide a connection for additional structural units—for example, vertical steel plates that will secure posts for an elevated deck or the anchor bolts that you use to secure a sill. Placement is important since the attachment must be located accurately. Even when installing strap-hinge supports in a masonry wall (perhaps for a gate), you want to be careful

Fig. 22–1 Inserts may be placed in the joints of masonry units and brickwork. Predrilled, steel plates made it easy to hang this wrought iron bracket after the masonry work was done.

Fig. 22–2 This heavy steel bar is imbedded in a concrete pour and then used to secure the post for a wooden stairway. Such bars should be U-shaped or have a right-angle bend at the bottom.

FIG. 22–3   Many types of special brackets are available to hold additional structural elements to a concrete pour. This U-shaped design is typical of those made to secure square posts.

about vertical alignment and spacing. The more time you take to be accurate, the easier the next step will be.

There are many types of add-on fasteners available today. When you are puzzled about which to choose, consider the following factors. Is the wall solid or hollow? Walls of plaster, Sheetrock, wood paneling, and cored masonry units

FIG. 22–4   Items may be added after the concrete has hardened. In this case, large holes were formed and then filled tightly with an epoxy patching compound after the rail posts were installed.

FIG. 22–5   Sometimes, the manufacturer of a product supplies special holding devices. The rings
may be placed anywhere in the channel. This permits some flexibility in placement
of the pipe.

are hollow. Poured concrete and brick are typical solid walls. For hollow walls,
you want a fastener that passes through the face and becomes secure when part
of the fastener spreads and grips against the back surface (blind side) of the
wall. For solid walls, you want a fastener that grips because it expands in the
hole that you drill for it.

How heavy is the attachment and what type of load will it put on the fastener? "Light" and "heavy" are relative terms, but small pictures and plaques,
pinup lamps, bud vases, and the like are probably light loads, whereas wall-hung
cabinets and wrought iron fixtures are heavy loads. Even the popular picture
hook which attaches to a wall by means of an angled brad or nail comes in different sizes; the weight that the hook will support is called out on the package.
You only need to weigh what you wish to hang. If one hook won't do the job,
use two. Theoretically, two hooks should support twice as much weight as one
hook. The point, though, is that too much can't hurt. Too little is always a
failure.

The type of load describes how the attachment tends to pull the fastener
from the wall. A straight-down force is a shear load, which is exerted by attach-
ments like pictures and mirrors. The fastener must be strong to resist breaking,
but the in-wall bond is not as critical as it is when the load exerts an outward
pull. A typical example is a wall bracket that supports a hanging planter. The
weight of the planter is down, but the support arm acts as a lever to pull the

STRENGTH FACTORS OF VARIOUS TYPES OF CONCRETE AND MASONRY IN RELATION TO ANCHORING DEVICES

Material	Strength	Remarks
Poured concrete	Excellent	Assuming a good mix
Block (concrete)	Good	Be aware of core sections that are too thin to take long anchors
Block (cinder)	Acceptable	Cinder block is softer than concrete block
Brick	Good	Locate anchors in mortar joints
Brick (veneer)	Good	Locate anchors in mortar joints; okay for moderate loads
Stone (veneer)	Acceptable	Load capacities similar to brick veneer
Stucco	Poor	Light loads okay; get more strength by penetrating into wall studs
Ceramic tile Marble Terrazzo	Acceptable	Very careful drilling is required; the material is usually thin and brittle; general rule is to plan for light loads only

FIG. 22–6 The lower post was secured with lag screws driven in the wood border-pieces. Lag screws were used on the upper posts too, but in expansion shields that were placed in holes drilled through the pavement.

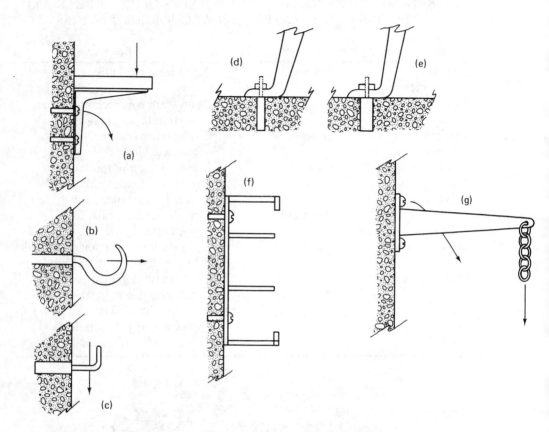

FIG. 22–7 Types of loads to consider in relation to fasteners: (a) load on a shelf is DOWN but tendency is for fixture to pull away from the wall; (b) hooks for items like clotheslines tend to pull OUT of the wall; (c) DOWN load on hooks for hanging items; (d) a workbench puts little load on the fasteners; (e) a machine bench sets up considerable vibration and so applies load in all directions; (f) cabinet load is similar to that of the shelf; (g) the longer the support arm the stronger the DOWN and OUT loads.

bracket away from the wall. This, of course, tends to pull the fastener out of the wall.

Static loads (dead loads) have a constant effect; you don't have to worry about changes in weight or sudden impacts. Other installations may have to resist constant or intermittent shocks and vibration such as those imposed by air conditioning units and power tools that are secured to a floor. The latter call for heavier fasteners and, ideally, fasteners that can be retightened when necessary.

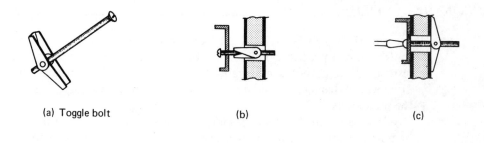

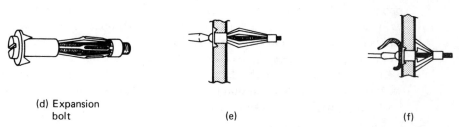

FIG. 22–8   Two common types of fasteners for hollow walls: (a) Toggle bolt—requires hole large enough for wings to pass through; (b) the bolt passes through the fixture and then through the wings; (c) the wings grip on the blind side of the wall; (d) Expansion bolt; (e) after the unit is inserted in drilled hole, the screw is turned to expand the split sleeve against the back side of the wall; (f) the screw is removed, passed through the fixture, and then turned back into the unit.

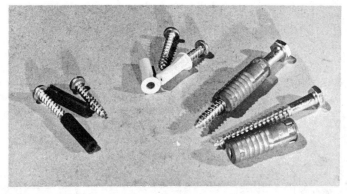

FIG. 22–9   Typical fasteners for solid walls. From left to right are simple fiber sleeves, special plastic sleeves with anti-turn fins, metal expansion shields. All come in different lengths and diameters.

FORMING HOLES

Most of the fasteners that you are likely to work with in either hollow or solid walls require a predrilled hole. The size of the hole must equal the size of the fastener. It must be exactly right if the bond is to have maximum strength. Usually, the diameter of the hole is mentioned on the package or is stamped on the product itself.

You can form holes in masonry by working with a star drill and a heavy hammer, but the system is rather passé today because of the availability of good portable electric drills and special carbide-tipped bits. Working by hand requires time and patience and a lot of care to keep the hole from becoming oversized. If you choose to, or must, work by hand, you can be more precise by drilling an undersized hole first. This simply means, for example, that when the job calls for a ½-in. hole, you use a ¼-in. star drill to begin with. Then you repeat the operation with a ½-in. star drill. Hold the drill erect and rotate it as you strike with the hammer. Don't use a cabinetmaker's hammer for such work; a heavy ball-pein hammer, even a light sledge is better. To facilitate such jobs, you can work with a special impact tool designed for use with star drills. Remember that a lot of light taps with the hammer will do a better job than a few heavy whacks. Be careful that you don't damage your hands while you are forming the hole. As always, safety goggles are a good idea.

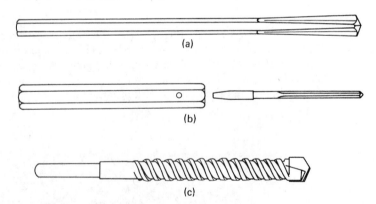

FIG. 22–10 To form holes, you can work by hand with a Star drill (a), which is available in sizes up to 1 inch, or with a smaller version, often called a Rawl drill (b) for holes under ¼ inch in diameter. The easiest way to work is with carbide tipped bits (c) made for use with electric drills.

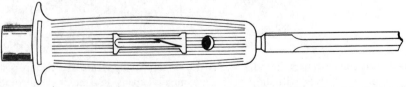

FIG. 22–11 Hand impact tool made for use with Star drills; it provides for a better, safer grip.

Fig. 22–12 One half inch drills with variable speed controls are available today. The power plus speed selection makes them good tools for general masonry drilling as well as routine drilling chores.

An electric drill that drives carbide-tipped bits makes it comparatively easy to form holes in masonry. A combination of power and slow speed is required, which just about rules out a single-speed ¼-in. drill for such work. The latter can be used successfully on wood paneling and even on Sheetrock and plaster but not on dense masonry. A better choice is a ⅜-in. drill, especially if it has

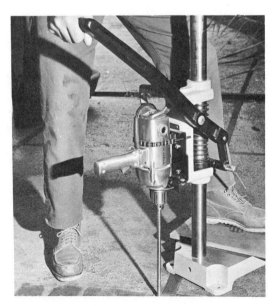

Fig. 22–13 For super accuracy and minimum fatigue, a portable drill may be mounted on a drill stand and used as shown here. This is a heavy-duty outfit but smaller tools can be used in similar fashion.

variable speed. The best choice is a ½-in. drill: It will have good power and rpm's for all the masonry drilling you are likely to encounter.

As with star drills, it is a good idea to work up to the hole size you need. If you go from ¼ to ⅜ to ½ in., you'll be taxing the drill less, prolonging the life of the cutting edges on the bits, and will get a neater, more precise hole. Don't be too cautious with feed pressure. The bit must cut consistently. If you are too gentle, the bit will do nothing but burnish. This will not form holes, and it will dull cutting edges quickly.

If you find that the drill you are using tends to stall or overheat, back it out frequently and allow it to cool. Masonry drilling doesn't require lubrication, but you may find that a very small amount of water in the hole will help to keep the bit cool.

Drilling speed will vary. You'll find the going slow if the bit should center on a large piece of aggregate. In such a situation, a few hard whacks with a hammer and a small piece of steel rod may fracture the stone enough so that the bit can pass through more easily.

DRIVING WITH A HAMMER

There are some types of nails that are specially hardened for driving into masonry materials. Generally, such nails are used for attaching a plate to a floor or furring to a masonry wall. You'll find that the degree of difficulty in driving them relates to the age of the concrete. When a severe problem is encountered, it is best to go to a fastener that can be used in a drilled hole. They are not meant to be used in a weight-holding capacity.

These nails have a degree of brittleness that can cause them to break when they are not hit correctly. A snapped nail can jump with considerable force, so take precautions to protect yourself: Wear gloves, safety goggles, and the like, and don't work with an audience nearby.

Use a good-sized hammer to drive such nails; the cabinetmaker's version won't do. A heavy, ball-pein hammer is acceptable; a light sledge is better.

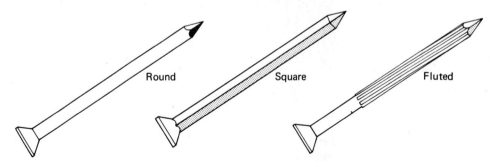

FIG. 22–14 Specially hardened nails for masonry and concrete come in the above shapes and are available in lengths from ½ to 3¾ inches and in #5 to #10 gauge. Fluted nails come in one large gauge size.

FIG. 22–15 Driving hardened masonry nails calls for some precautions. Use a heavy hammer that is in good condition. Wear safety goggles. The Hand Tools Institute also recommends a hard hat.

EXAMPLES OF FASTENERS

It would be impossible to show every type of fastener made, but we will illustrate an assortment that have wide application and that are usually available from local supply sources. Your best bet is to check out large hardware stores, the hardware section of department stores, and local homeowner's supply centers. Chances are good that when the item you want is not on display, the dealer can order it for you.

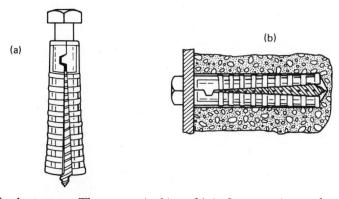

(a)

(b)

FIG. 22–16 Shields for lag screws. They come in ¼ to ¾ inch screw sizes and require a hole that equals the outside diameter of the shield (a). Driving the screw causes the shield to expand in the hole (b). NOTES: Such items come in "long" and "short" in each size. Use the long for more strength in a weak structure. When using in brick, try to position in a mortar joint rather than the brick itself.

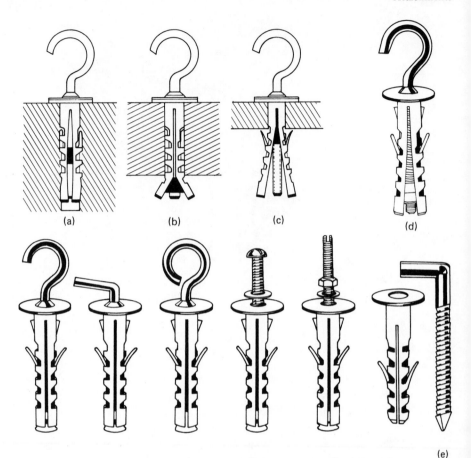

(a)   (b)   (c)   (d)

(e)

FIG. 22–17 Fischer SB Nylon anchors are available with hook, eye, screw, or stud. All are 1¾ inch long and require a ⁵⁄₁₆ inch hole. They can be used in solid material (a), hollow core material (b), and thin material (c). Turning the screw draws an expander cone into the anchor (d) thereby creating pressure against the sides of the hole. The hook shown in (e) is driven with a hammer.

FIG. 22–18 This new Molly Parabolt is a "drop-in," high strength, concrete anchor that is available in 1¾-inch and 3¼-inch lengths. Both sizes have a ¼-inch diameter.

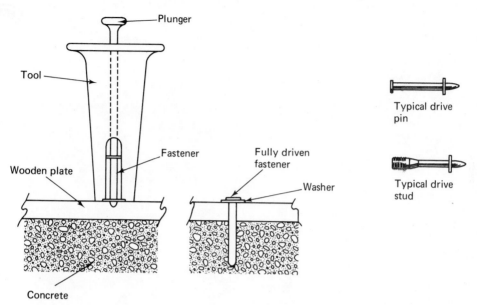

Fig. 22–19 Special fastening devices are installed with a tool made for the purpose. The tool is gripped in the hand and the plunger is struck with a light sledge hammer to drive the fastener. The example shows a wooden plate being secured to concrete. The washer is part of the fastener. It slides up as the fastener is driven home. Typical drive pins range in diameters from $\frac{1}{8}$ to $\frac{3}{16}$ inch and in lengths from $\frac{1}{2}$ to 3 inches; typical drive stud sizes are 8–32, 10–24, $\frac{1}{4}$ inch—20, and lengths range from $1\frac{3}{8}$ to 2 inches. NOTE: Such fasteners are most useful in concrete. They are not recommended for brick, tile, or brittle materials.

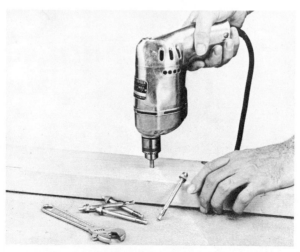

Fig. 22–20 The hole for the Parabolt fastener may be drilled directly through the material to be attached. The hole size must match the diameter of the fastener.

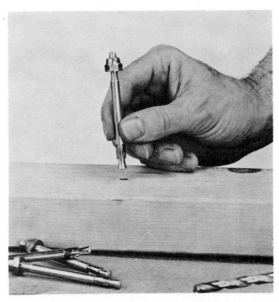

FIG. 22–21 The anchor is dropped into place, or tapped very gently with a hammer to seat it. Taking up on the nut causes an expansion in the sleeve area so the fastener snugs up in the hole.

FIG. 22–22 The smaller lengths are good for applications such as this. Follow instructions on the package that suggest correct hole depths for various applications.

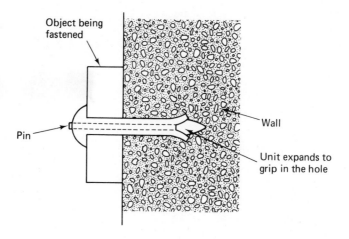

Fɪɢ. 22–23 Pin-Grip fastener (Star) is inserted in a hole drilled through the fixture and into the wall. Driving the pin with a hammer expands the unit to grip in the hole.

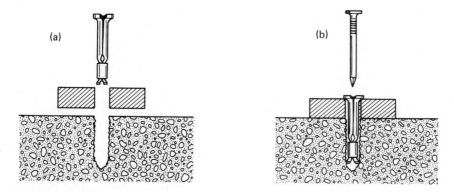

Fɪɢ. 22–24 Type of expansion sleeve anchor (Star's "Dryvin") you can use with nails: (a) hole required equals the outside diameter of the sleeve and is drilled through the object to be fastened and into the wall; (b) driving the nail expands the sleeve to grip in the hole.

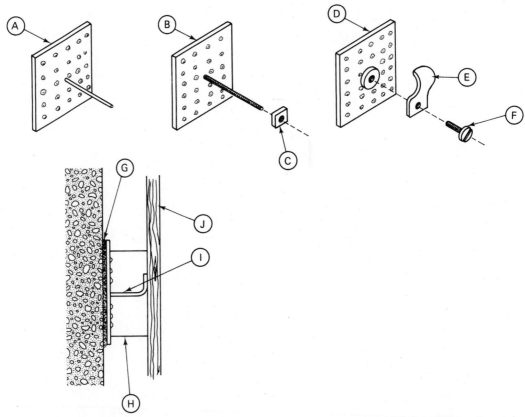

FIG. 22–25 Examples of adhesive anchors—these are stuck to the wall with special mastics: (a) nail type; (b) bolt type; (c) nut; (d) hanger type; (e) support that may be used for pipe or conduit; (f) screw. Example of how nail type may be used to attach furring strips for wall paneling; (g) adhesive; (h) furring strip is impaled on nail and the (i) nail is clinched; (j) the wall paneling is then secured to the furring strips.

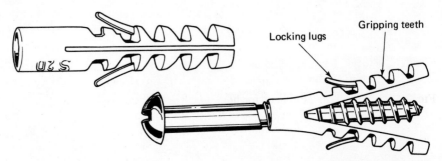

FIG. 22–26 S-type nylon anchors (Fischer) will work with wood screws, lag bolts, even nails. Different sizes are available to take screws that range from #2 up to a ⅝ inch lag. NOTE: The length of the screw should equal the thickness of the workpiece plus the length of the anchor.

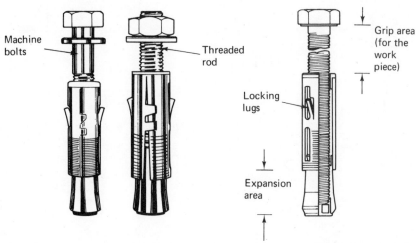

FIG. 22–27 Heavy duty steel anchors identified as Fischer S/L will work with machine bolts or threaded rods.

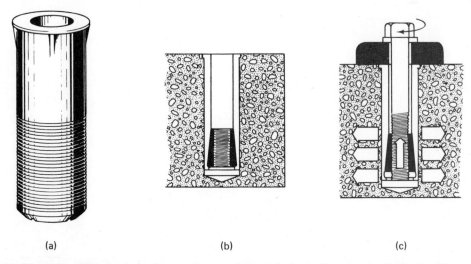

(a)   (b)   (c)

FIG. 22–28 M anchors (Fischer) for heavy duty machine bolt fastening: (a) available for $3/16$ to $5/8$ inch machine bolts—anchor lengths range from $1 3/8$ to $3 5/8$ inch; (b) once installed, the anchor provides a reusable thread in masonry; (c) turning the bolt causes a built-in brass cone to draw up into the shell—the action creates expansion and a strong bond in the depth of the hole.

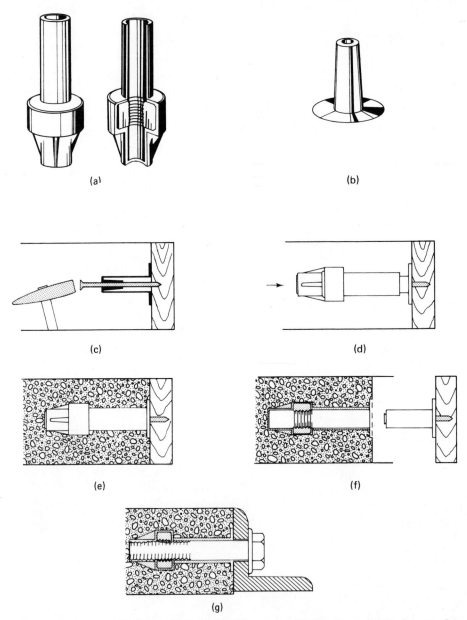

Fɪɢ. 22–29 B-M type inserts (Fischer) are imbedded in a concrete pour to provide for attaching with machine bolts. The inserts are available for machine bolt sizes from $\frac{5}{16}$ x 18 up to 1 x 8 (a). They are used with matching stud plates (b) and the stud plate is nailed to the wooden form (c). The insert is slipped over the stud plate (d) and the concrete is poured (e). The stud plate removes with the wooden form (f), thus providing a permanent, threaded insert.

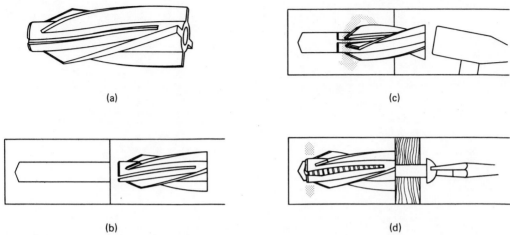

(a)

(c)

(b)

(d)

FIG. 22–30 GS (Fischer) anchor is a special design for lightweight (or foam) concrete: (a) sizes are available for #10–16 and $\frac{5}{16}$ to $\frac{3}{8}$ inch screws; (b) hole size should be one-half the outside diameter of the anchor; (c) the wings thread into the concrete as the anchor is driven home; (d) the anchor expands as the screw is set.

QUESTIONS

22–1   Name some attachment materials that are included *during* a construction job.
22–2   Name some attachment materials that may be added *after* a construction job.
22–3   Describe a typical fastener for a hollow wall and one for a solid wall.
22–4   What is a *Star drill*?
22–5   What kind of drilling tools can be used with electric drills to form holes in masonry materials?
22–6   What is the best drill size to use when drilling holes in masonry, and why?
22–7   What are good safety precautions to take when you are driving special masonry nails?

23

Pillars and Posts

The area of pillars and posts runs the gamut from simple wooden fence posts to impressive masonry columns carefully spaced before an entryway. All pillars and posts should be considered an important part of the landscaping theme since they affect reaction to the scene as a whole. Jobs that are well done and designed for appearance as much as function reflect a love of craftsmanship and respect for the entire surrounding area.

Installation and construction procedures for pillars and posts should not be approached in casual fashion. Fence posts that are misaligned and weakly installed will fall or, at least, tilt and become eyesores. Even small masonry pillars are quite heavy, so they require footings that are substantial enough for the job on hand.

Plan in advance for units that can do more than support fence rails. You may wish to include, for example, outdoor lights, studs for gate hinges or bolts for wooden frames, brackets or support arms for house-number plaques or a mailbox, and so on. Usually it is easier—often it is imperative—to provide for such additions during the initial construction. It is quite simple to include a center conduit as part of a masonry post project, or to provide a passageway for electrical cable in a wooden post before it is installed. As an afterthought, it can be a problem.

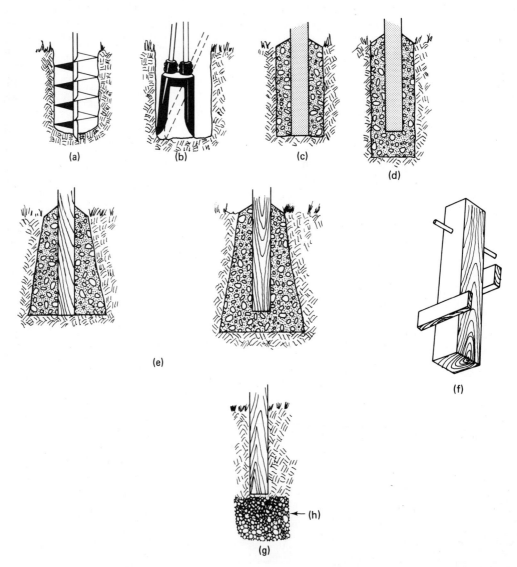

FIG. 23–1 How to set fence posts. (a) Form a 6 inch diameter hole by working with hand or powered auger or by (b) using a clam shell digger. Avoid working with a spade since it will make too large a hole. (c) Pour the concrete after the post has been set plumb. (d) Post ends may be encased to provide more protection. Set posts a minimum of 18 inches to 24 inches below grade. (e) You can increase strength if you work with a hand spade to key the hole after it is formed. (f) Small cross arms (wood or steel) can be attached to the imbedded area to increase rigidity. (g) Light-duty posts can be set directly in soil by drilling a hole and then back-filling with the post in place. (h) It is a good idea to have a 6 inch gravel bed in the bottom of the hole. NOTE: For the greatest protection, creosote the imbedded end of the post. Slope the top of the concrete for drainage.

WOOD POSTS

It is not a good idea to use a shovel to dig holes for fence posts. Since wooden posts are usually 4 × 4's, the diameter of the hole doesn't have to be more than 6 in. This applies whether you plan to backfill with soil or fill with concrete. Larger holes require excessive amounts of concrete and disturb too much soil. Digging the minimum-sized hole with a shovel is a tough assignment but easy to do with a clam-shell digger or an auger. Both these items are standard tools that you can buy or rent. The auger is available as a hand tool or it may be powered. The latter makes sense when you have many holes to dig. When you investigate, check out different models. Some can be handled by one man, others require two operators.

Dig the holes about 6 in. deeper than the burial length of the posts so that the bottom of the hole can have a subbase of loose gravel to provide for drainage. Since fence posts are usually of redwood or some other insect- and rot-resistant material, they are often placed in the hole as is. However, if you believe very strongly in safety precautions, you'll want to paint the buried portion with a preservative regardless of whether you embed in soil or concrete.

The post sits right on the gravel subbase. If you use a rather stiff concrete mix, its possible to keep the post plumb without temporary bracing. This can also be accomplished when you backfill with soil. Don't, however, neglect bracing just because it is a nuisance. If you need it, use it. Some professionals will ac-

FIG. 23–2   The clam shell digger has jaws that close when the handles are pulled apart. The handles are held together when the tool is thrust into the soil.

FIG. 23–3   To use auger, bear down on it heavily and then twist. The blades are designed to cut as they turn and to collect loose soil so it can be removed by pulling up the auger.

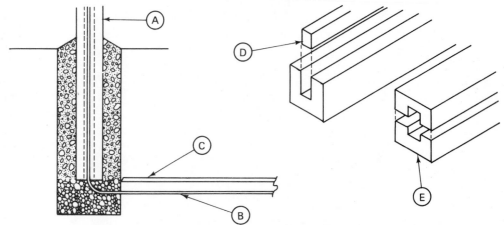

FIG. 23–4   Running wire through a wooden lamp post. The wooden post is shown in (a) and the special underground cable in (b). If the code does not permit the special underground cable follow local standards for buried wire. The redwood strip (c) protects the wire. Make an opening through the post by cutting a deep groove and then filling part way with a wooden strip (d), or by cutting matching grooves in similar pieces and then joining the pieces with waterproof glue (e).

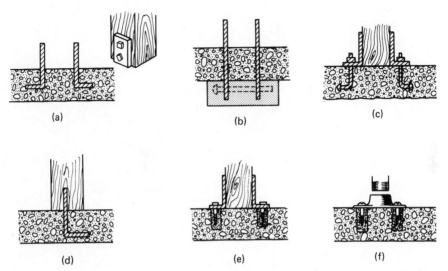

(a)   (b)   (c)

(d)   (e)   (f)

Fig. 23-5   Various methods of securing wood posts to concrete slabs. Steel angles are imbedded in the concrete during the pour, and are spaced to receive the posts (a). The post is fastened with lag screws or through bolts. Steel bars (b) are set in a precast pier that is set below the pour. Bent-over bolts are imbedded in the pour (c) and used to secure steel angles. Reinforcement rods (d) are bent and imbedded in the pour. The bottom of the post is drilled to slip over the rod. Expansion sleeves (e) may be used in holes drilled in the concrete after it has set. The idea may be used to secure angles or (f) pipe flanges when the post is standard pipe.

tually construct the fence without first securing the posts in the holes. The theory here is that the togetherness of all components—rails, sheathing, and so on—assures plumbness in all areas. It does make sense when circumstances permit it. A straight fence on a level grade is easier to do this way than one that must be stepped down in sections to conform to a slope.

Soil backfill should be done in slightly dampened layers that are well compacted. Place concrete with a shovel and use a piece of 1 × 2 to do a minimum amount of tamping. Place enough concrete so that you can form an above-grade slope on either side of the post. This will help to direct water away from the post's foundation. Allow the concrete to set for about 24 hours before proceeding to the next steps.

CAST CONCRETE

For our purposes, you can regard concrete as a viscous material that flows to fill all areas of a form made to produce a particular shape. So, in addition to substantial slab and foundation projects, you can create posts of various types with comparatively small cross sections. The posts can be for mailboxes, number plaques, outdoor lighting, and the like, as well as for heavier units that may be combined with wood for fences and screens.

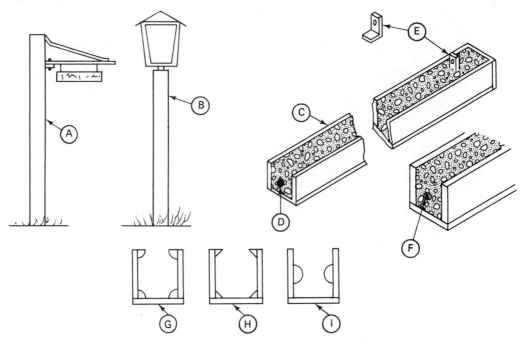

Fig. 23–6 Concrete posts for house numbers (a) and outdoor lights (b). Make a form for post height plus length to be set below ground in concrete—18 to 24 inches (c). Use ½ inch reinforcement rods for solid posts (d), and set in hangers (e) if needed. A conduit or pipe for electric cables (f) takes the place of a reinforcement rod. Add corner details by using quarter-round molding (g) or triangular strips (h). You can also do side details (i).

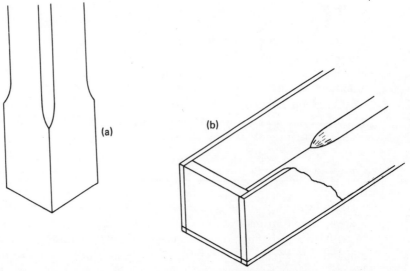

Fig. 23–7 To shape a cast concrete post this way (a), use triangular or one-quarter round strips with ends rounded off (b).

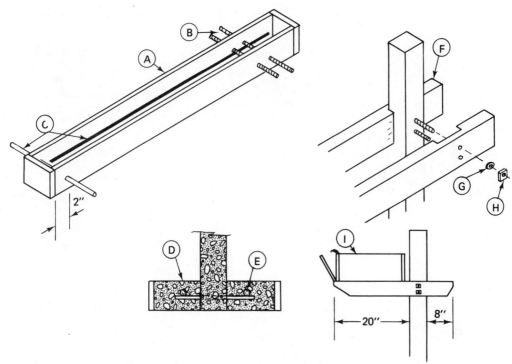

FIG. 23–8   Concrete mailbox with its own base. Make a form to cast a post 4 inches x 4 inches
x about 5 feet (a). Insert ⅜ inch threaded rods through the form—space them
about 2 inches apart and 14 inches down from the top (b). Insert a ½ inch reinforcement rod through the form and 2 pieces of ⅜ inch reinforcement rod down
the center. Wire-tie the latter to the ½ inch rod and to the threaded rod (c). Make
a form for the base that will measure 4 inches x 24 inches x 24 inches (d). Install
pieces of reinforcement rod to strengthen the base (e). Cross arms (f) are 2 x 4s
notched to fit the post. They are secured with washers (g) and nuts (h). Cut off
the excess threaded rod. The mail box (i) is secured to the 2 x 4 wooden cross arms.

When the cross section of the project is on the small side, use a 1- to 2- to
2¼-part mix with a ½-in. maximum aggregate size. When the project has
greater bulk—see, for example, the fence posts shown in the sketch—stay with a
conventional mix and a ¾-in. (or more) maximum aggregate size. To get a result that is pure white without having to use paint, buy white portland cement
and white aggregates.

Make the forms carefully but use a minimum amount so that stripping will
be easier. Be sure that all concrete-contact surfaces are either oiled or coated
with a release material. To be safe, allow the casting to remain in the forms for
about 48 hours. Then, water-cure for at least a week. Select a work site so that
the project will be in the shade during the curing period.

Slender posts can be designed with parallel sides, or they may be tapered;
corners can be square or shaped with special insert material in the forms. Even
the sides of posts can be detailed. The few ideas that we show here in the

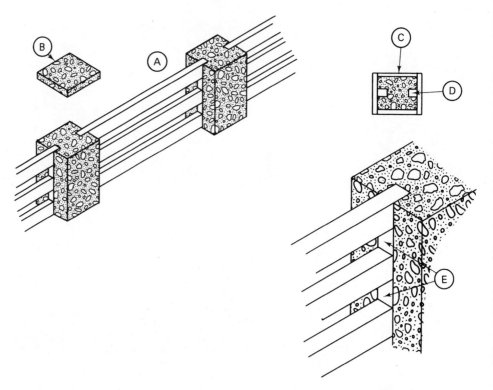

FIG. 23–9 Concrete fence posts: cast-in-place posts grooved for the fence boards (a); precast cap (b); section through the form needed for the posts (c); key blocks which provide the grooves (d). For spaced boards instead of a solid fence, use wooden spacer blocks (e).

sketches are only examples of the kind of things you can do. Other, similar, cast-concrete procedures and projects will be shown in Chapter 24.

BRICK AND BLOCK

Posts, piers, and pillars constructed of masonry units require adequate footings just like any full-sized project. When the top surface of the footing will be level with or below the grade, you can cut the forms directly in the soil. Be sure that the dirt walls are vertical or that they slant out a bit at the base. Take all necessary subbase and subgrade precautions.

When the footing surface will be above grade, you can often get by with a shallow above-grade form used in combination with a cavity in the soil. Often, a dirt wall is sufficient to hold the wooden part of the form in place. If not, you can brace it with bricks or blocks.

Now is the time to think about any vertical steel reinforcement or conduit for electrical cable. Pipe for cable should be bent out at the lower end in a

Fɪɢ. 23–10 Brick posts blend nicely with wood components and with greenery. Note the light on the post. Planning for such things should be done before construction starts.

Fɪɢ. 23–11 Be sure that the base for the footing is firm and level. Check the surface of the wood section with a level. You don't want to compensate for errors here by having to do tapering mortar joints.

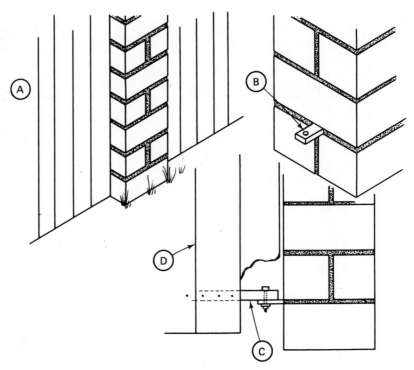

FIG. 23–12 Fence posts of concrete block (a); upper and lower steel ties set into mortar joints (b); fence rails bolted to steel ties (c); and fence boards nailed to the rails (d).

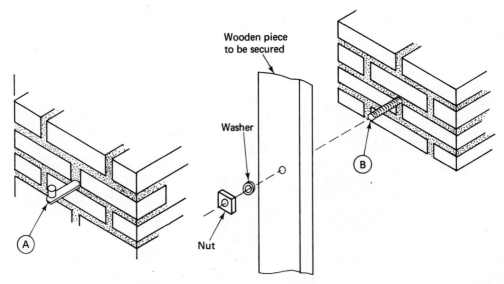

FIG. 23–13 Two ways to provide for a gate (or an arbor) between masonry posts: (a) install a hook for a strap hinge in mortar joint; (b) provide for a wooden frame between posts by installing bolts in mortar joints.

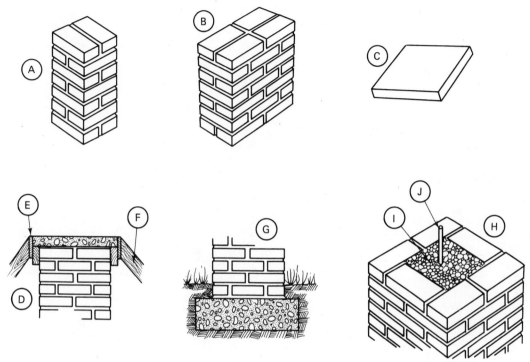

FIG. 23–14 Brick posts (or piers). The bricks in each course determine the size and shape: (a) 2-brick; (b) 4-brick. The top can be capped with a precast slab that you buy or make (c), or you can construct a wood form and pour on the job (d). The form is two pieces of wood nailed together (e) and braced from the ground (f). All such projects should be on solid footings that are below the frost line—check local codes (g). You can set brick to form a box (h) that you then fill with rubble and concrete (i). Pipe is used for electric cable when the post is for a light (j).

sweeping curve, especially if the cable will have to be snaked through after the masonry is up. Vertical steel may be driven into the soil with a sledge hammer and the concrete poured around it. Large posts may be left hollow, or they may be filled with rubble and grout as the walls are built up.

Be sure that you plan well for any items that will be attached to the posts. These may include openings for rails, predrilled, projecting steel bars for attaching fence supports, hooks for strap hinges, jambs for hanging gates, and so on. Usually, the hardware can be built in as the construction goes on. Be careful about vertical alignment and spacing.

Use the same construction techniques that you would employ if you were building a full wall. Use a level frequently to check the horizontal plane of each course and the plumbness of each wall. It is also a good idea to use a large square as you go to be sure that corners are square.

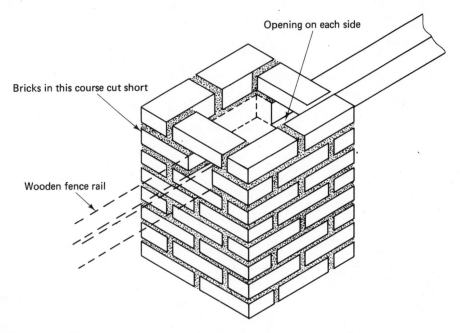

FIG. 23–15 Brick post with opening for fence rail. Use a precast cap to cover the post.

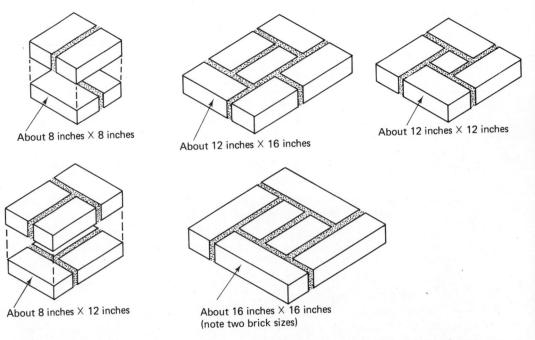

FIG. 23–16 Examples of solid masonry piers. When the project must support a heavy, concentrated load, its height should not exceed 10 times the smallest dimension.

Second, fourth, sixth, etc. courses

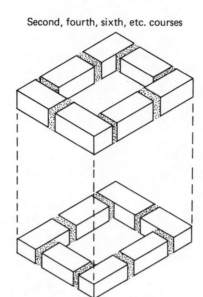

First, third, fifth, etc. courses

FIG. 23–17 How to build up a hollow brick pier (or post). Check frequently with a square and
·a level to keep work square and plumb. Bricks placed as shown will make a pier about
16 inches x 20 inches.

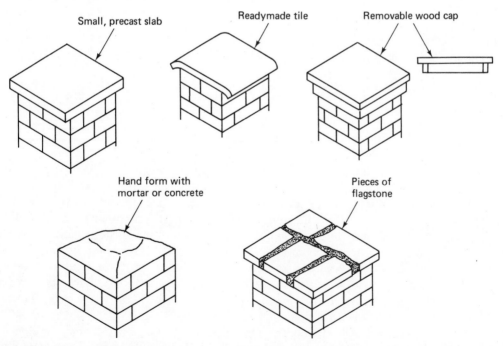

FIG. 23–18 Topping masonry posts. In all cases, a slight slope to shed water is a good idea.

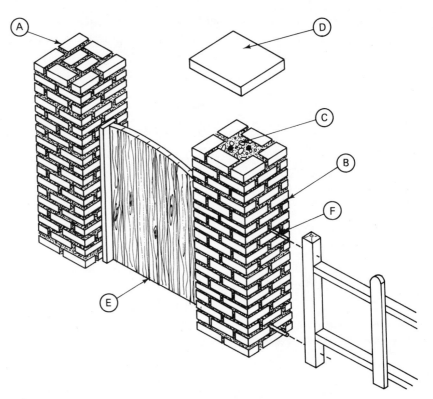

FIG. 23–19 Masonry gate posts. For solid brick masonry (a) or hollow filled with concrete (b) use ½ inch reinforcement rods set in the post footings (c). A precast slab may be used as a cap (d). The gate hangs on jamb pieces bolted to the masonry (e). The fence post bolts are imbedded in the masonry as the job progresses (f).

PORTABLE POSTS

You can make removable or portable posts by working with heavy-gauge steel tubing and a concrete base. This type of arrangement works well for tetherball or clothesline posts, or for posts that will support badminton or volleyball nets. A socket is placed in the ground to hold the post secure when it is needed.

An easy way to do the job is to sink a 5-gal drum in the ground. Set it deep enough so that its top surface will be just a bit below grade. Be sure that the base it sits on is firm and level: Dampen it and tamp it; use sand or gravel as a fill if necessary. The sleeve you use should provide a slip-fit for the post. Set both the sleeve and the post in place before you pour the concrete. This will make it easy to check plumbness of the post with a level.

You can make a wooden plug or use a suitable piece of large dowel to seal the sleeve when the post is not being used. This will keep the sleeve from filling

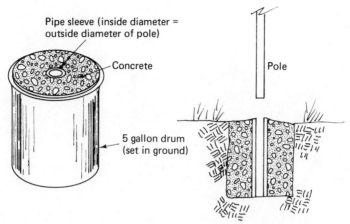

Pipe sleeve (inside diameter =
outside diameter of pole)

Concrete

Pole

5 gallon drum
(set in ground)

FIG. 23–20 One way you can organize for a portable clothes pole or anything that is similar. The same thing can be done without the 5 gallon drum simply by digging a hole in the ground. In either case, set the pole and the pipe sleeve in place before pouring the concrete. This is done so that you can use a level against the pole to check for vertical alignment.

with dirt or water. Do be sure that the plug doesn't project above grade so that it will not be a tripping hazard.

Another approach is to use a 5-gal drum but to set the post directly in the concrete. To move the post, you tilt it and roll it about on the bottom edge of the drum. An old standby is to use a discarded automobile tire as a permanent,

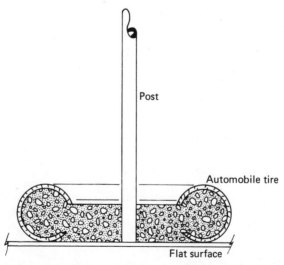

Post

Automobile tire

Flat surface

FIG. 23–21 This is an old trick but it works as a portable post. Place a discarded automobile tire on a flat surface. Place and brace the post in plumb position—then pack the inside of the tire with concrete.

FIG. 23–22 Above-grade supports for posts can be cast in many shapes. Even though a base of this nature is heavy, it may be considered a portable project since it is not imbedded in the ground.

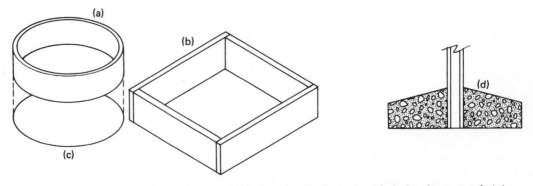

FIG. 23–23 Above-ground bases for posts can be round—formed with light sheet metal (a) or square—form made of wood (b). Either can have a bottom, or you can do without by setting the form on level ground for the pour (c). It is a good idea to slope the top for drainage (d).

FIG. 23–24 Supports for add-on wooden posts can be supplied as shown here. The height of the concrete pier was established for levelness with the concrete floor that will be poured later.

FIG. 23–25 Pier forms can be lengths of heavy cardboard tubing that is sold for the purpose. Note the brackets imbedded in the piers. These will support posts for an outside deck.

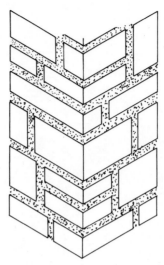

FIG. 23–26 Cut stones have a degree of uniformity that makes it possible to lay them almost like bricks. Careful selection of pieces for each course results in strong, staggered joints.

above-grade form. Place the tire on firm, level ground or on a sheet of plywood. Center the post or sleeve and then pour concrete around it. Be sure that you tamp enough to fill the tire completely and that you use enough concrete so that you can slope the top surface. This will supply ample rigidity for the post, yet it will be easy to move the post if you tip it over and roll it about on the tire.

QUESTIONS

23–1   Why is it unwise to dig holes for fence posts by working with a shovel?
23–2   What two tools are commonly used to form holes for fence posts? Why?
23–3   Make a cross-section drawing of an installed 4 in. × 4 in. fence post.
23–4   What should you do when you want a white concrete post and do not wish to paint it?
23–5   Make a sketch of a form that can be used to cast a concrete mail-box post. Show dimensions, reinforcement, etc.
23–6   Make a sketch that shows a hollow brick post (or column) and its foundation.
23–7   Describe two readymade forms that can be used to make portable posts.

24

Concrete Crafting

It may be difficult to change your point of view of concrete from the substantial slabs and walls that we have talked about to projects that are purely decorative and strictly for fun; yet the material *is* that flexible. Concrete (or mortar) is not like a clay that you can hand-form in the round, but it is fluid enough to be shaped by what you use to contain it while it hardens.

As an example of imaginative use, consider that there are craftsmen who stuff a heavy balloon or even a paper or burlap sack with a stiff mortar mix and then work the material to achieve a particular shape. The rubber, paper, or cloth is stripped off after the mix has set sufficiently. In a sense, the material is used as a sculpture medium.

For some projects, such as tables and benches, a conventional mix with steel reinforcement rods or wire mesh is required. For others, such as plaques or masks—projects that need only self-sustaining strength—you can use a lightweight mix.

A typical lightweight-concrete formula substitutes vermiculite for standard aggregates. A ratio of 1 part of cement to 5 parts of vermiculite is acceptable but on the borderline as far as strength is concerned. To reduce weight without sacrificing strength, use a mix that is 1 part of cement to 2 parts of sand to 3 parts of vermiculite. Pumice or haydite may be used in place of the vermiculite. Combine 3 parts of one of these materials with 2 parts of sand and 1 part of

FIG. 24–1 The kind of concrete benches you often see in parks and shopping malls are not difficult to cast. The form you need is just a case with inside dimensions that equal the outside dimensions of the project.

FIG. 24–2 Commercially produced planters are done in special double-wall forms that are hinged or sectioned to open. You can produce the same kind of project by using sand as the mold material.

FIG. 24–3 These plaques are set up in a heavy wood frame so the whole becomes a screen or fence. The plaques were cast on a sand bed. The raised details were achieved by impressing the sand.

cement for a general-purpose material. To get a more plastic mix with any of these materials add about ¼ part of hydrated lime or fire clay.

Dry colors may be added to the mix, but because the amounts needed may vary from brand to brand, read the instructions on the package and obey them as if you were working with a conventional mix.

Since many craft projects fall in nonstructural categories, some personal experimentation is permissible. Following are some of the things that have been tried with an adequate degree of success. In all cases, the improvisation consisted of using a strange material in the mix.

As a substitute aggregate, use sawdust or wood chips, crushed walnut shells, styrofoam fill, marbles and glass beads, and the like. This may sound rather daring after all the admonition about doing the mixes exactly right, but we're not building houses or walks now.

Some of the castings you come up with may be soft enough to carve after they have been removed from the forms. Mixtures that contain a lot of vermiculite, for example, will be much easier to carve than those that are predominantly cement and sand. Some craftspeople have used plaster instead of cement with vermiculite. This produces a white casting that is flecked with the color of the vermiculite.

Vermiculite is probably available locally either as an agricultural aid or an insulating material. The type used for agricultural purposes has small particles and is stocked by nurseries and garden supply stores. That used for insulating has larger particles and should be available from building supply dealers and maybe large lumberyards. Even a large department store that has a home maintenance section might stock the material or be able to get it for you through a catalog.

Forms for most of these projects can be made of wood or sand. Sometimes the two methods are used together. If you plan work in this area, have a half yard or a full yard of sand dumped in a pile at your worksite. Then you can impress the sand freehand or use a preassembled wood shape to form the cavity for what you wish to cast.

PLANTERS

Concrete is a good material for planters because it retains moisture, thereby keeping soil from drying out. The weight of the project depends on the thickness of the walls and the mix that you use. Walls that are less than 1 in. thick are tough to do unless you preform a sheet of wire mesh into the shape you want and then coat it with a heavy-duty mortar mix as if you were doing parging.

Planters cast in wooden forms are pretty strong when the wall thickness is more than L in., but, in any event, you can provide additional safety by including wire mesh or steel bar reinforcement. Because of thin cross sections and limited space, all reinforcements should be preformed very carefully and placed

Fig. 24-4   Planters may be done with a special mix for an exposed aggregate finish. This calls for removal from the sand-mold as soon as possible—but carefully—so the project may be brushed and washed.

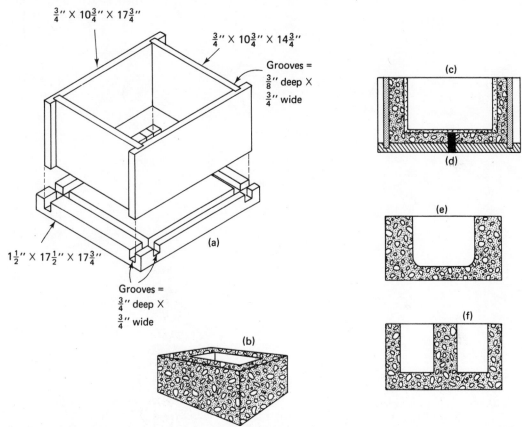

$\frac{3}{4}'' \times 10\frac{3}{4}'' \times 17\frac{3}{4}''$

$\frac{3}{4}'' \times 10\frac{3}{4}'' \times 14\frac{3}{4}''$

Grooves = $\frac{3}{8}''$ deep X $\frac{3}{4}''$ wide

(c)

(d)

(a)

$1\frac{1}{2}'' \times 17\frac{1}{2}'' \times 17\frac{3}{4}''$

Grooves = $\frac{3}{4}''$ deep X $\frac{3}{4}''$ wide

(e)

(b)

(f)

FIG. 24–5   How to make a reusable form for concrete planters: (a) all parts mate in grooves—interior surfaces are coated with oil or a release before each casting—sides and ends are held tightly together with strong cord; (b) finished casting; (c) inside form can be a sand-filled cardboard box; (d) greased dowel to provide drain hole; (e) inside shape can be achieved with objects like wastebaskets, large cans, etc.; (f) use two cans to make separate cavities. Forms are removed after the concrete sets. Clean form parts immediately and re-oil before each use.

very accurately. You want to be sure that the bars of the mesh are centered in the wall thickness.

Wooden forms can be made simply for one-time use, or you can design them for reuse. Forms that have hinged sides or that are joined temporarily by means of a groove system (see sketch), are typical "production" tools. Parts should be thoroughly cleaned immediately after the casting is removed and re-oiled or coated with a form release before the next pour.

Whatever craft project you tackle, be sure to give it plenty of time to cure before you apply stress. Remove the casting from the form very carefully. Place it in the shade on some simple platform that will provide air space beneath.

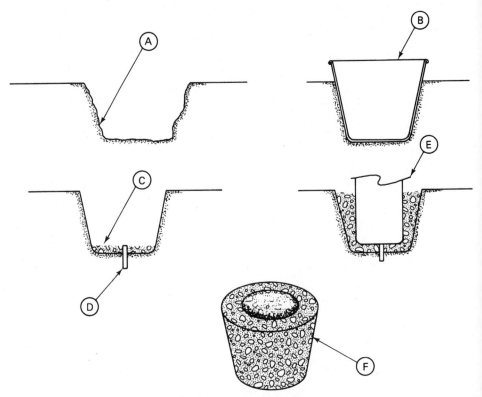

Fig. 24–6 Sand casting a round planter. Scoop out an approximate cavity in a pile of damp sand (a). Use a wastebasket or some similar item to finish-form the cavity (b). Pour some of the mix to form the base (c) and use a greased or oiled dowel to form the drain hole. Mix pouring is completed around an inner form—oil-coated can, tube, etc. (e) and your planter is completed (f).

Cover the project with burlap or similar material or even a thick layer of newspapers. Keep the covering wet for at least a week.

To demonstrate a typical procedure, we will now describe how to cast a planter that has straight sides and square corners. Use plywood to make a form whose inside dimensions are those of the planter you are making. Drill a ½-in. hole through the center of the bottom and then plug it with an oiled dowel that is long enough to penetrate the thickness of the casting. The purpose of the dowel is to provide a drain hole through the casting; more than one can be included if you wish.

Have an inner form ready. This, too, can be a wooden box that you make or you can work with a ready-made cardboard box that you fill solidly with sand as the pouring is done. Casting is done by placing the base layer of concrete, adding the inner form, and then filling in around it. The mix can be a bit looser than you would make it ordinarily but, even here, don't overdo. Use a slim piece of wood as a tamper and tap around the form with a hammer as the pouring progresses.

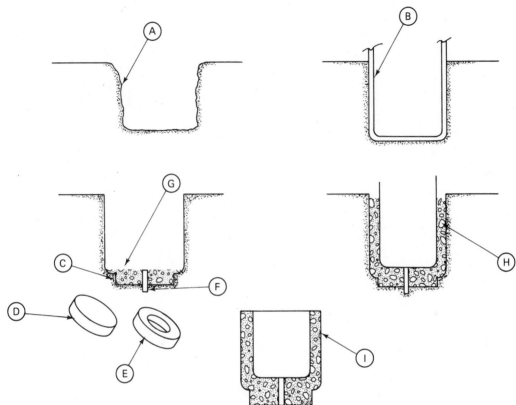

FIG. 24–7   Sand casting a round planter with integral footing. Scoop out an approximate cavity in a pile of damp sand (a). Use a drum or something similar to finish-form the cavity (b). Form the footing by pressing in (c) a solid disc (d) or a ring (e). A greased or oiled dowel is used for a drain hole (f). Pour enough of the mix to form the bottom of the planter (g). Then work around an oiled can or tube to finish the pour (h). (i) is a section through the completed planter.

The kind of wood you use for the outside form will affect texture. Smooth wood equals smooth texture; etched plywood, for example, will result in a wood-grain effect. For exposed aggregate, remove the forms as soon as the concrete has set enough to hold together and then work the surfaces with a whisk broom and water. If the concrete is stubborn, you can work with a wire brush. For special exposed aggregate effects, use small-sized colored stones in the mix.

Round planters, as we show in the sketches, can be cast by using a pile of sand as the outside form. The sand should be damp enough so as to compact and hold together when you squeeze a fistful.

Scoop out an approximate-sized hole in the sand and then use a bucket, a wastebasket, a drum, or whatever, to finish forming the cavity. Work to compact the sand firmly around it and then twist the container as you bear down on it. To remove the forming implement, twist it as you lift. If the sand is in the right condition, you will have a well-formed cavity for the pour. You now follow

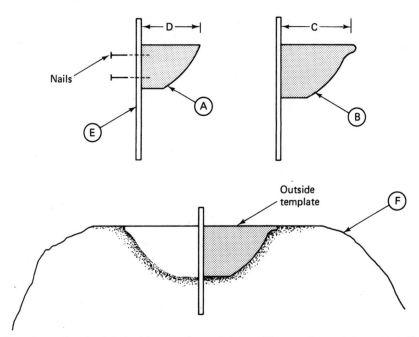

FIG. 24–8   Sand casting a bowl: (a) inside template; (b) outside template; (c) greater than (d) by the thickness of the casting; (e) nail ¾ inch templates to ¾ or 1 inch D dowel; (f) sandpile with leveled top. Use outside template first to form bowl cavity in the sand. Use the inside template to form the bowl as you pour the concrete. The dowel, which is used as a pivot, also forms a drain hole in the bottom of the bowl.

the same steps that were outlined for casting in a box. Work carefully so that the sand mold will hold. The inner form can be a similar but smaller container —a cylinder of some sort, a square box, whatever you wish to use that will provide adequate wall thickness. When the concrete has set sufficiently, you can scrape the sand away from the outside so that the planter can be removed, or you can wet-cure it while it remains in the sand.

Bowls can be cast in sand molds but they require special templates (see sketch) to achieve correct shapes. Be sure that the sand is damp enough to compact firmly. Press the dowel into the sand, and slowly rotate the outside template. Remove excess sand as you go. If you do this job carefully, you'll get a smooth bowl shape. If cavities appear, just plaster them with damp sand and use the template again.

Begin the pour after the inside template is in place. Add concrete or mortar carefully and in small amounts that can be spread easily by the template. If necessary, tamp spots lightly with something like a rubber ball. Fill voids with small amounts of the mix.

This kind of project can be accomplished with a concrete mix that has minimum-sized aggregates or with a strong mortar mix. After you are through shaping, let the pour set until it is hard enough to resist marring; then cover the

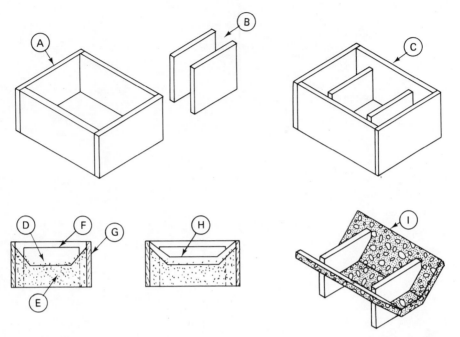

Fig. 24–9   Cast a planter with integral wooden legs. Make a bottomless box as a mold (a). Cut two legs—redwood is preferable (b). The legs are set in the mold and damp sand is used to complete the form as shown in the cross section (c). Shown are small, galvanized nails in the legs (d), sand (e), the legs (f), and the box (g). Stiff mortar or concrete with very small aggregate is used for the casting (h). The finished planter is shown in (i). Use ¾ inch stock as leg material for small planters and 2 inch stock for larger ones.

whole thing carefully with damp sand. It won't hurt to let the project sit for a couple of days, wetting the sand occasionally with a fine spray from a garden hose.

Don't yank the bowl when the time comes to remove it. Instead, scrape the sand away and lift the project from the bottom. Then, wet-cure it for about a week.

PLAQUES

Concrete plaques should appeal to the eye—something like the attraction that prompts you to adopt a special piece of driftwood or stone. You can hang it on a wall, place it on a table as is, or combine it with other materials to create a conversation piece. The reason for its existence doesn't have to be anything more than an urge to create.

Installations range from simple wall-hanging, through partial burial as a decorative detail in a garden scheme, to framed inserts in a fence or screen. You can cast with regular or lightweight concrete, create geometric or free-form

FIG. 24–10 Pierced plaques are as simple to do as solid ones but they do call for extra form work. These were set up between heavy beams that were well established in concrete. Pierced designs appear light and airy.

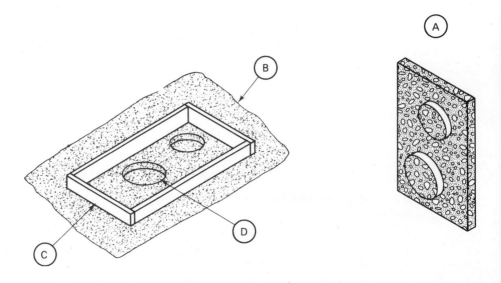

FIG. 24–11 Example of a raised plaque. The backing (a) should be at least 1 inch thick. The sand bed (b) should be level and firm. The outside form may be above the sand bed or imbedded in it (c). The forms for raised areas are pressed into the sand (d).

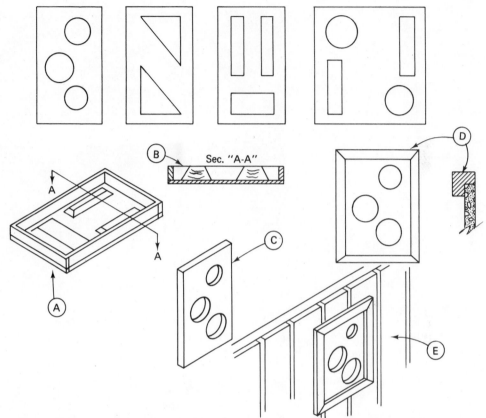

Fig. 24–12 Examples of pierced plaques. A typical form of 1 inch material with plywood base is shown in (a). The size of the plaques may vary—the thickness may range from 1 inch to 2 inches, depending on the size and casting material (see text). Beveled edges on inserts (b) will make cast removal easier. It also contributes to the appearance of the casting. Such plaques may be placed as-is (c) or framed with wood for hanging (d) or for an inset in a fence (e).

shapes, have pierced or solid results, incise or raise the surface—the potentials are as limited as your imagination.

Forms can be made of wood or sand or you can combine the two. For example, set a rectangular, bottomless form on a bed of compacted sand. Use a kitchen spoon, or something similar, to form a design in the sand within the borders of the wooden frame. Install a hanger (see sketch) as you pour the concrete, and when the project has cured, you'll have a ready-to-hang wall plaque.

Another way to get raised details on plaques is as follows: Cut a sheet of ½-in. or ¾-in. plywood to the size of the plaque you wish. Use a keyhole saw or a saber saw (or work on a jigsaw) to cut openings in the plywood. The shape of the openings can be square, round, fish or plant form, letter or number. Just be sure that the sides of the cuts are square or, better still, beveled a bit toward what will be the inside of the form. Frame the panel with border strips of 1 × 2

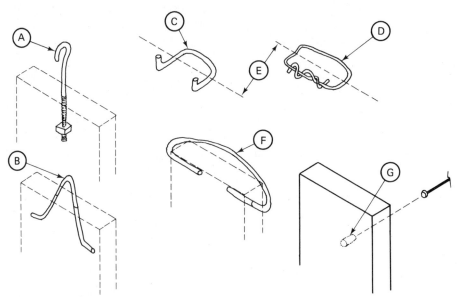

FIG. 24–13 Methods of hanging plaques: (a) eyebolt or hook; (b) bent up rod; (c) bent up rod emerging from back of casting; (d) heavy wire emerging from back of casting; (e) back surface of casting; (f) bent up rod; (g) use greased dowel to form blind hole so casting can hang on heavy nail or bolt.

FIG. 24–14 The family plaque has prints of all members including the dog. Making the prints is just a matter of pressing down on the concrete before it has hardened. Scratch in the year with a sharp stick.

wood and then set the whole thing, panel side down, on a level surface of compact sand or on another sheet of plywood or hardboard. Pour slowly to be sure that the cutouts are filled and then screed so that the pour will be level with the upper edges of the frame. Let the concrete set for about 48 hours and then turn the form over very carefully to remove it from the casting. As always, don't forget to oil the forms before you do the casting.

You can hand-print plaques by pressing down on the mix before it hardens, or by pressing your hand into a bed of damp sand to create a mold for casting. The former makes an incised print; the latter, a raised one. Be careful when you press *into* sand and when you remove your hand. You may have to try several times before you get a mold that looks good enough to cast, but doing it over is just a matter of releveling the sand. If the mold you form has slight imperfections, you may be able to smooth them by working with the convex side of a small spoon or the eraser end of a pencil.

BENCHES AND TABLES

Benches and tables are somewhat more challenging than planters and plaques but only because they are bigger and their structure therefore requires greater

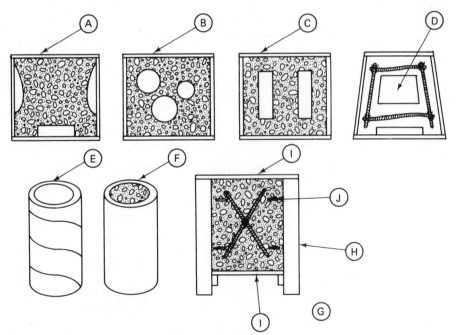

FIG. 24–15 Some ways to cast concrete legs for garden benches: (a) shaped wood blocks to create form; (b) various size cans to form holes—be sure to oil the cans; (c) inset bricks; (d) wood blocks for openings; (e) large mailing tube or winding core as a form; (f) filled tile—round or rectangular; (g) combine wood and concrete; (h) 4 inch x 4 inch redwood legs; (i) forms removed after concrete sets; (j) spikes.

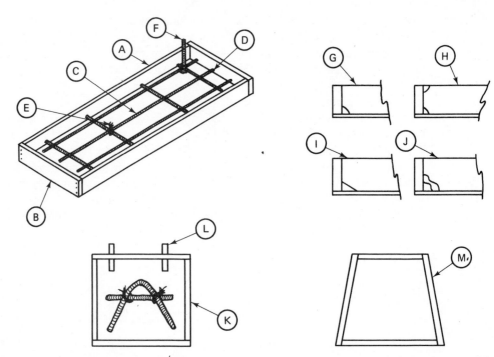

F<small>IG</small>. 24–16 Basic ideas for casting concrete garden benches. Use 1 x material to make a form
with *inside* dimensions of 4 inches x 20 inches x 5 to 6 feet (a). Use ⅝ inch
plywood for a base (b). You can do without a base if you place the open form on a
plastic sheet on a level work surface. Use ⅜ inch reinforcement rods as shown in
(c). Keep the rod ends about 1 inch short of the forms (d). Use wire ties at each
intersection (e). Situate ½ inch x 5 inch bolts in four places, 5 inches in from the
ends and 5 inches in from the sides (f). Note that the rods and bolt heads should
be midway in the pour. To get a fancy edge you can use one-quarter round molding
(g). This can be done for both top and bottom edges (h). Or you can use a triangular strip (i) or even fancy molding (j). The leg forms should be dimensioned so that
the total width of the casting is 2 inches less than the bench top and the total bench
height is 17 or 18 inches (k). Use greased ½ inch dowels or bolts to form sockets
for the permanent bolts in the slab (l). The legs may be tapered (m).

attention. Consider all benches and tables to be composed of slabs and legs.
They may be all concrete, cast in one piece, or the slab and legs may be cast
separately and then joined with hardware or even an epoxy cement. You can
combine a concrete top with wooden or wrought iron legs, or make concrete
legs for a wooden slab. Other masonry materials, such as brick and block, can
be erected as piers to support either a concrete slab or a wood surface.

The form for a one-piece concrete bench or table is made like a tight-fitting
case. All the *inside* dimensions should equal the *outside* dimensions of the
project. The form doesn't have to be pretty but it must be sturdy and assembled
with easy stripping in mind. This calls for butt joints, placed so that each is
accessible for separation. What you must do is picture the stripping operation

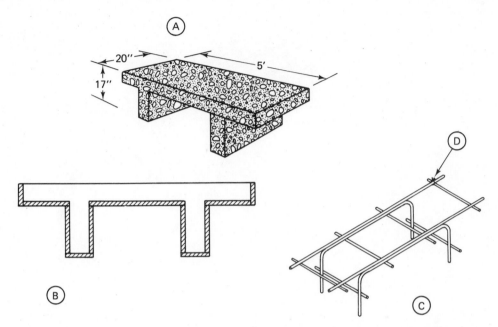

Fig. 24–17 One-piece concrete bench. The dimensions shown in (a) are fairly typical. The thickness of the top and legs should be 3 inches. A section through the form required to cast the bench is shown in (b). Careful removal of the parts after the concrete sets will permit their reuse. Reinforcement rods of ¼ or ⅜ inch are preassembled as shown in (c). Use wire ties (d) at all intersections.

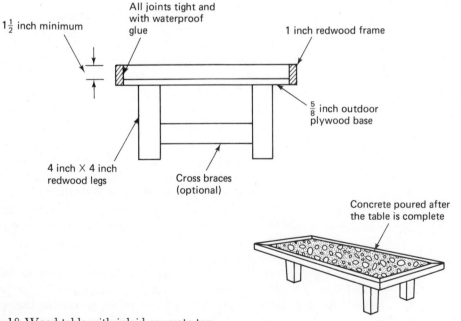

Fig. 24–18 Wood table with inlaid concrete top.

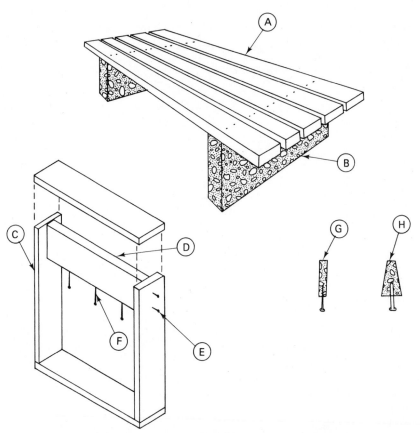

FIG. 24–19 How to provide for nailing a wooden bench top to concrete legs: (a) wooden seat boards; (b) cast concrete legs; (c) form for the legs; (d) nailing strip becomes part of the pour; (e) drive two nails on each side just deep enough to hold the nailer; (f) 16d galvanized nails to lock nailer to the concrete; (g) nailer can be straight board; (h) keyed nailer provides better joint.

in a sequence that exposes the joints of adjacent form members each time you remove a piece. The reinforcement for such a project is truly a preassembled skeleton that is laid out so that it will sit midway in the pour in all sections. Rod ends should not be closer than 1 in. to perimeter lines of the project.

Coat all concrete-contact surfaces with a release and set the form on a level surface, preferably in a shady place. Use a regular concrete mix with a maximum aggregate size of either ½ or ¾ in. Start filling in the legs, tamping enough to assure settling the mix in the form. Place the steel skeleton when the legs are about one fourth full. If you wish, you can keep the skeleton elevated in the slab area by using a few small pieces of stone as temporary platforms. Continue to pour carefully so that you don't knock the reinforcement out of position. Tamp as you go, and tap frequently with a hammer on the outside of the forms.

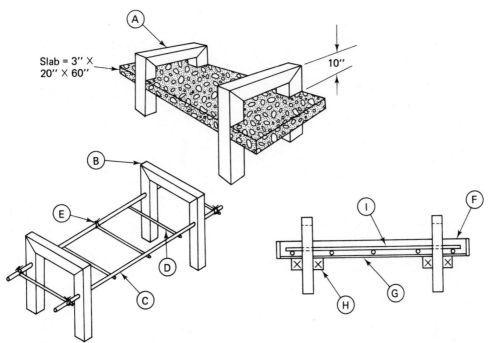

Slab = 3" X
20" X 60"

10"

FIG. 24–20 Bench with integral wooden arms. Use 4 inch x 4 inch redwood (a). Make a tight
miter joint—use waterproof glue (b). Galvanized pipe of ⅝ inch or ¾ inch reinforcement rods pass through holes drilled in legs (c). Reinforcement rods of ¼ or ⅜
inch are used (d), and wire ties are used at all intersections (e). A temporary form
for casting the bench top is shown in (f). The base of the form is notched for the
legs (g). Temporary supports for the form are shown in (h). Reinforcement material is used midway in the pour (i).

Screed the top when the form is full and, after a bit, finish off with a float. You
can if you wish, use an edging tool to get round corners on the surface.

If you don't plan a special finish, let the pour stay in the form for a couple
of days. Keep the whole thing covered with any material that will help retain
moisture. After the second day, tilt the project very gently to its side, and then
again so that the legs are uppermost. Let it sit in this position for another day
or so and then remove the forms. You may be able to remove the form in one
piece by having one person at each end tugging gently upward. If the form
doesn't move rather easily, work by disassembling it. Either way, work so that
there will be minimum shock to the casting. Go through the usual wet-cure
procedures.

You'll have to remove the forms sooner if you wish to give the casting special surface treatments, but do wait until the concrete has hardened enough to
hold together. Also, be extremely cautious when you remove the forms.

As the sketches show, there are other ways to make tables and benches. A

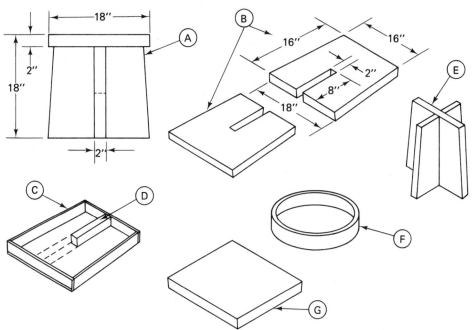

Fig. 24–21 Small patio table. The table is cast in three pieces (a). The legs are dimensioned as shown in (b). Note that a slot is at the top of one leg and at the bottom of the other. A form for casting the legs is shown in (c). A block is used to form the slot (d). Use in the position shown for one leg and in the dotted line position for the second leg. (e) shows how the legs fit together. A form for the top is made by wrapping a thin sheet of metal around a plywood disc (f). The table top may also be square (g).

Fig. 24–22 Benches (or tables) can be heavy wooden slabs supported by masonry piers. The piers can be brick or block or even poured concrete. Whatever, be sure they have adequate footings.

popular method is to build a wooden table with a framed top that is actually a form for a pour. When all the woodwork is complete, the top is filled with mortar or concrete, which is then finished off by floating. With this project, any kind of finishing can be done—brooming, brooming and washing, seeding with special aggregates, and so on.

Before you pour, stud the inside surfaces of the top frame with galvanized nails. Lay down a piece of wire mesh midway in the pour. Since the woodwork is part of the project, it should be assembled with well-made, tight joints and bonded with a waterproof glue.

Study the sketches for ideas that show how you nail wood seats to concrete legs and how you can design so that wood–concrete combinations may be done as one-piece projects.

WATERFALLS AND POOLS

The sound and the sight of moving water have a special fascination that can contribute much to any garden. Projects in this area can be quite large and elaborate enough to serve as the main point of interest or they can be scaled down to act only as decorative details. Designs can be executed with any kind of masonry or natural materials, or the two can be combined very effectively.

FIG. 24–23 The pond for this waterfall project is a concrete shell with a natural stone coping. The pump is hidden beneath the rocks of the waterfall. Lightweight lava stones were used.

You can be quite formal with what you do or construct so naturally that the project appears to have been lifted in toto from a cool, forest glade.

At one time, a garden waterfall project required a constant supply of water. The water was drained off from a lower pool and directed to a drainage area or reused for irrigation purposes. This limited enthusiasm for the inclusion of a waterfall in a landscape theme. Today, because of the availability and the low cost of submersible, recirculating pumps, waterfalls can be included anywhere, even *inside* the house, without concern for the cost of water or its waste.

The pumps function by sucking water from a pool and then raising it to a higher level from where it cascades back to the pool. A plastic hose routes the water from the pump to the upper level where it can emerge as a solid stream, or empty into a basin and overflow to areas below. The pumps come in various sizes, so selections can be made in relation to the volume of water to be moved and the distance it must rise. A typical small pump will move 120 gal of water per hour to a height of 5 ft or 200 gal per hour to a height of 1 ft. A larger version will move 300 gal of water per hour to a height of 5 ft or 430 gal of water per hour to a height of 1 ft. In all cases, the higher you pump, the less water you can move. Regardless of the pump you use, you can install a regulator or a restrictor on the plastic hose to control the flow of water to the rate you desire. This permits adjustments after construction so that the waterfall will be just right for the effect you want.

Water that falls vertically from a great height makes more noise than water

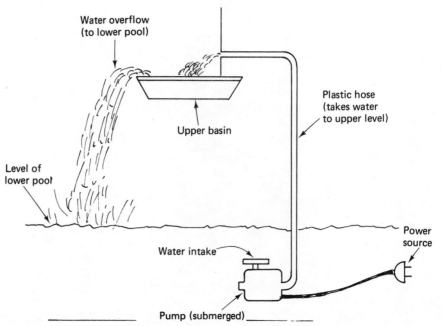

FIG. 24–24 The basics of a recirculating water pump.

flowing down a slope or a series of ledges. This must be considered when you design. The sound of water in the background is pleasant when it is subdued. But if you must raise your voice to be heard above it, it can be very annoying. This is especially critical when the waterfall is on a patio with nearby conversational seating. If you find, after construction, that the noise level is too high, you can change it by restricting the water flow and by introducing breaks in the path of the water.

Waterfall or pool apparatus may be purchased as a kit that includes the pump, some decorative materials, and a large sheet of plastic. The latter item is used as a pool liner so that in-soil work can be limited to digging a cavity and lining it with the plastic. Then, all you'll need is a coping of natural stones and a platform of some sort for the water to fall from.

You can construct fountains as well as waterfalls if you install a spray nozzle on the pump. The nozzles are usually called "fountainheads" and are designed for different effects. A single spout version will throw up a column of water; an adjustable head with few or many jets will result in multiple sprays; a ring with perimeter jets will produce a circle of sprays. It is also possible to create water patterns. You can, for example, have the pump push water into a sealed copper tube in which you have drilled a series of small holes.

When you design a fountain, you must be sure that the pool is large enough (in area) to receive the falling water. Controls that are on the fountain heads and methods of restricting water flow allow you to decide how big the pool should be.

Some kits consist of precast fountains and waterfalls that you simply set in

FIG. 24–25 This small project was done with a kit that includes a plastic liner to seal a cavity that is dug in the soil. Statuary may be used as shown here, or it may be placed directly in the pool.

FIG. 24–26 This interesting fountain was done by casting a concrete basin and then installing a concrete bowl. The water spurts up in a single stream that fills the bowl and then overflows to the basin.

FIG. 24–27 Plug-in type waterfalls are simply set in place and decorated. Some have built-in lights and pockets for small plants. This type of product may be used indoors or out.

FIG. 24–28 Fountains are done with the same recirculating pumps. Special nozzles direct the water upward in a single column or a number of individual sprays. Restrictors help to control water height and volume.

FIG. 24–29 Here, an old crock at the upper level receives the water from the pump and drops it to a basin where it overflows into the pond. Be sure to design so that water-noise doesn't become a nuisance.

place and plug in. Most of these are made from fiberglass materials, realistically shaped and textured. Some have built-in lights and pockets for small plantings. They may be used indoors or out, as is, or as the core of a larger, more elaborate project.

When you design from scratch, see the project as a two-level affair. The lower level is a pool, the upper level is the maximum height to which you can, or choose to, raise the water. Between the two levels can be a shear drop, a smooth or interrupted slope, a series of ledges that in themselves can be small pools, even a system of small buckets or pots that overflow into each other.

The pool can be of concrete that is poured to the level of the grade, or higher by doing some above-grade forming. Waterproof paint may be applied after the concrete has cured. You may use concrete as the base of the pool and then build up walls of brick, block, or even stone. For these materials, you apply a parge coat to just above the water level and then finish with a waterproof paint.

You can think in similar fashion about what you wish to do above-grade. Most garden waterfalls are built up of natural stones, but ideas can range from a masonry wall that is veneered with ceramic tile to wooden buckets or clay pots that hang from frames. When the upper-level design is open, for esthetic reasons you can substitute copper pipe for the plastic hose that is usually used.

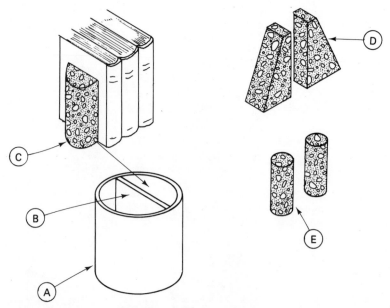

Fig. 24–30 Ideas for bookends: (a) a 2 or 3 pound coffee tin, or something similar; (b) a tight-fitting wood divider—the tin may be cut away after concrete or mortar sets; (c) glue felt to the bottom of such projects; (d) cast block that is cut on the diagonal; (e) cylinders cast in mailing tubes. The mix for projects like this can be colored, or the project can be painted after it dries.

MISCELLANEOUS PROJECTS

There really isn't too much that you can't cast by using either regular or light-weight concrete. Think about making bookends, lamp bases, doorstops, paper-

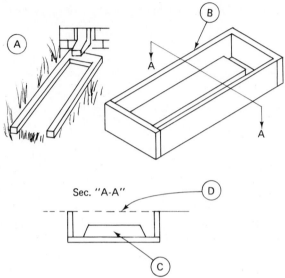

Fig. 24–31 A splash block at downspouts carries water away from the foundation wall. Such items are available ready made (a). To cast your own, make a long box about 4 inches deep and secure a trough block to the bottom (b). Bevel the edges of the block (c). The bottom of the project is shown in (d).

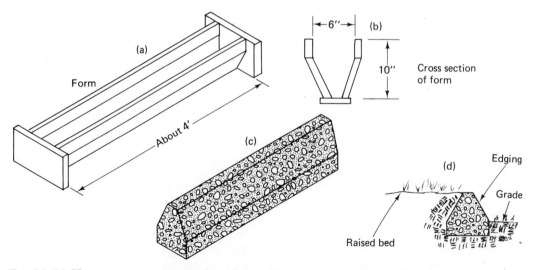

Fig. 24–32 How to cast concrete garden edging. Use a 1 x material to make a form (a). A cross section of the form is shown in (b). Dimensions may be changed to suit but do not cast the width less than 4 inches. (d) explains how the cast edging (c) may be used.

weights, water-runoff troughs, garden edging, special blocks to be used like brick, patio ashtrays, flowerpot stands, and shrines. Quite often, projects in this area can make use of concrete left over from a larger job.

Forming can be done with wood or in sand. When you allow yourself to think of concrete as a craft material, you will begin to see many packaging items and disposables as ready-made forms. Examples include small, cardboard boxes for bookends and paperweights, large mailing tubes or winding reels for lamp bases, tin cans for ashtrays and small planters, automobile tires for concrete

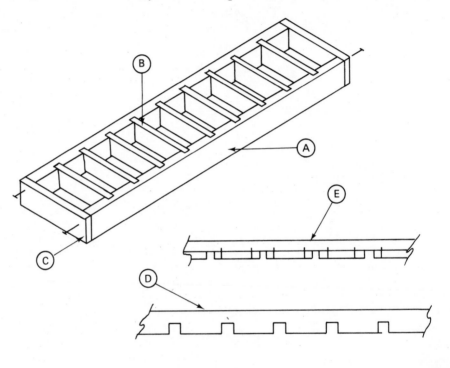

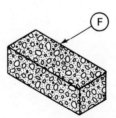

Fig. 24–33 Making a form to cast concrete blocks. The sideboards (a) are 2 x 4s and the dividers (b) are 1 x 4s. The ends may be tack-nailed (c). The divider grooves in the sideboards (d) may be cut on a table saw, or you can nail 1 x pieces together to accomplish the same thing (e). The form pieces should be dimensioned to produce blocks about 4 inches x 4 inches x 8 inches.

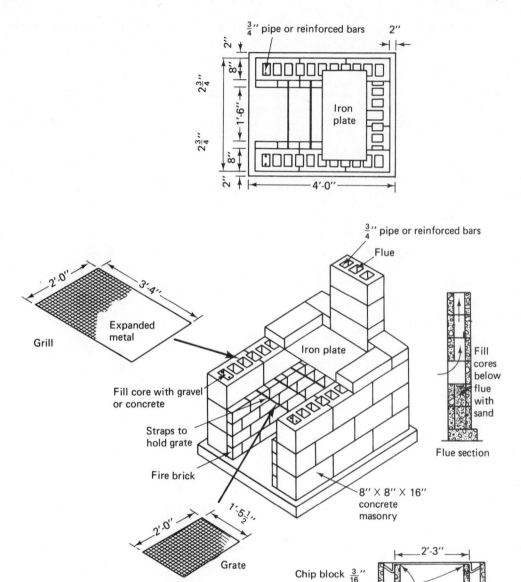

FIG. 24–34 Construction details of a typical outdoor fireplace (Courtesy of Portland Cement Association).

rings (you cut the tire away after the concrete hardens), odd pieces of conduit or pipe for casting cylinders, and so on.

When the item that inspires you isn't rigid enough for the job, you can always bury it in sand to get the support you need when pouring and tamping.

The items you make don't have to look like concrete unless you want them to. After a correct cure period, you can coat with paint or stain, even get metallic effects by applying, for example, a coat of plastic aluminum or steel.

QUESTIONS

24–1   What materials may be substituted for regular aggregates to get a lightweight concrete?

24–2   Describe a formula that will produce a strong, lightweight concrete.

24–3   How can you make such a mix more plastic?

24–4   What are some of the other "off-beat" materials that can be used in a concrete-crafts mix?

24–5   Make a cross-section sketch that shows how a bowl can be cast in a sand pile. Show *inside* and *outside* templates.

24–6   Design a concrete plaque and show the form that must be made for it.

24–7   Design a concrete slab that may be used as the top of a bench. Show dimensions, method of attachment to legs, reinforcement, etc.

24–8   Make a sketch that shows how wooden legs may be attached to a concrete bench or table top.

24–9   How does a recirculating pump work?

24–10   Make a cross-section sketch that shows how a recirculating pump works.

24–11   How can you regulate the flow of water from a recirculating pump?

24–12   Design a patio fountain. Make a rough sketch and call out materials and dimensions.

25

Maintenance

Major problems with masonry projects almost always occur if the job is not done right to begin with. The causes of isolated or general excessive settlement in concrete slabs, walls or posts that tilt, brick or flagstone walks that become tripping hazards, can be traced to rushed or haphazard preparation of subgrades and subbases. Maintenance then comes to include a substantial amount of repair which could have been avoided by preventive measures that accompany honest craftsmanship.

Time and weather do take a toll, but even here the effects can be minimized and countered by, for example, sealing joints between similar or dissimilar materials with caulking compounds and by repairing cracks in masonry joints and slabs before they become major problems. Many chores in these areas should not wait for the problem to appear. The wise worker will be his own inspector and will check periodically for possible trouble spots. Water damage should be watched for very carefully. It can enter between wood–masonry joints and cause rotting, fill spaces between flashing and masonry (for example, around a chimney), penetrate a veneer and appear on interior walls, freeze and cause serious ruptures. Prevention is easy; remedial tasks can be a headache.

CAULKING

The average homestead has countless lineal feet of joint lines that should be made weathertight. The easiest way to fill and seal the joints is to use a caulking compound that stays plastic while it develops a tough, surface layer. This permits a degree of movement without cracking while the skin opposes air and water penetration. Thus, a normal amount of movement in the joint members will not cause a reopening of the gap.

There are four ways to buy caulking materials: in a preloaded cartridge used with a gun that squeezes out the material through a nozzle; in bulk form for application with a putty knife or for loading a barrel-type gun; in rope form; and in small tubes where you squeeze out the material just as you do toothpaste.

The preloaded cartridge is probably the most popular method. Any kind of caulking can be purchased this way and in reasonable quantities; you can therefore have an assortment if necessary, without too much cash outlay. Since the materials will be in separate cartridges, you can go from one to another without fuss. The bulk form is useful when bought in small containers for jobs that call for application with a putty knife, but as a supply for barrel guns it is generally

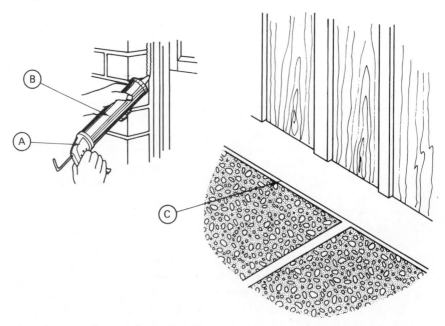

FIG. 25–1 Caulking. Gaps between masonry and framing can be sealed with commercial caulking compounds. A special gun (a) is used with a cartridge tube (b) that contains the caulking. The same method can be used to plug openings between, for example, patios and foundations. It is important to remember that the gap must be filled— laying a surface bead will not do the job.

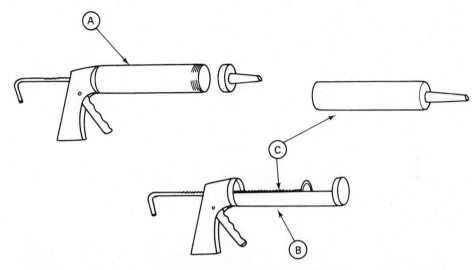

Fɪɢ. 25–2   Caulking guns: (a) used with caulking materials purchased in bulk—its barrel is filled with material taken from another container; (b) used with pre-loaded cartridges; (c) the cartridge fits in the half-barrel of the gun.

reserved for commercial use. Rope caulking is very convenient to use but is not available in all types of materials. Squeeze tubes are mostly for small jobs. Don't let price be *the* major factor when choosing a method or a material. Quite often, more money at the start buys a more efficient, more durable material, and this means less money spent in the long run.

CAULKING MATERIALS

Type	Use	Remarks
Oil base	Indoors or out; on wood, masonry, metal	Economical but not long-lived; takes paint
Silicone (rubber)	Anywhere; excellent for bathtub sealing	Costly but long-lasting; most types cannot be painted
Butyl (rubber)	Good adhesion between dissimilar materials	May be painted after suitable cure period
Latex base	Fast drying; acceptable for most caulking jobs, also usable as crack filler	Water-soluble; can be painted quickly (after about 30 minutes.)
Polysulfide (rubber)	Indoors or out; very good where much movement occurs in the joint	Usually requires careful surface preparation before application

*Be sure to read manufacturer's instructions on the container before using any caulking material.

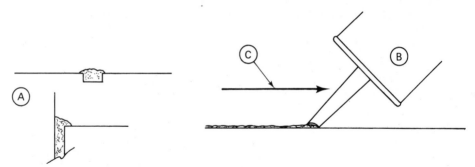

FIG. 25-3 Using a caulking gun. A good bead will fill the gap and overlap edges of the opening (a). Hold the gun at about a 45° angle (b). Lay down the bead in the direction indicated by the arrow (c).

Many of the materials require specific application procedures and a cure period before they should be covered with paint, so read the instructions on the package. Do be sure, regardless of what you select, that the application surfaces are clean and dry. The joint must be free of dust, dirt, moisture, flaking paint,

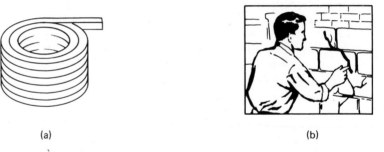

(a)   (b)

(c)   (d)

FIG. 25-4 Caulking may also be purchased in cord form in scored rolls so individual strands may be removed for use (a). It is simply pressed into place for such things as plugging small cracks in masonry (b), sealing around windows, doors, etc. (c), or for general weatherstripping chores (d). NOTE: always use such materials according to instructions printed on the package!

and the like if the caulking is to stick where it should and how it should. Some caulking compounds require a primer. If the package instructions call for one, you may be sure that it is needed.

Using a caulking gun efficiently requires a bit of practice. You must aim to correlate the amount of material you squeeze out with the speed at which you lay down the bead. When the gap is deep, you must squeeze harder and move more slowly than when the gap is shallow. It is somewhat tricky but nothing that can't be learned in short order.

Following is a list of typical places where a caulking seal can be very useful:

All joints between the house siding and framing members around such openings as windows and doors.

Under windows where the sill meets the siding; also at drip caps over windows and doors.

All flashing around chimneys, vents, and attic ventilators.

Where exterior masonry abuts the house—patio slabs, steps, and planters.

Between siding and mating members of any extension from the house, such as an entrance cover and carport.

Connecting corners of dormers, gables, and roof materials.

CRACKS IN CONCRETE

Cracks in concrete, even small areas that are worn or web-cracked, can be repaired relatively easily. Work with a hammer and chisel to form a patch area that has reasonably straight lines and a uniform depth of at least 1 in. Figure that what you are cutting away will be a form for the patch pour. Its walls should be undercut so that the new concrete will be keyed to the old and so that there will be no feathered edges that will, inevitably, flake and chip. This applies whether you are repairing a wide area or a long narrow crack.

Work with a brush, even a vacuum cleaner if you have to, to remove all loose materials and dust. Wet the area thoroughly, but don't leave standing

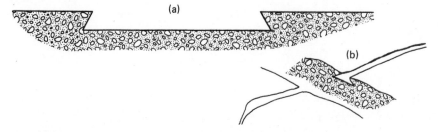

FIG. 25–5 When cleaning out a damaged area of concrete, always undercut the edges of the cavity (a). This applies to cracks as well as larger areas (b).

pools of water. Fill the cavity with new concrete, but strike it off so that it is a bit higher than adjacent areas. Allow it to set a bit and then float it to the true level. Follow curing procedures as you would for any pour. It will be difficult to hide completely the signs of repair, but you can minimize the impression with careful work.

There are special cement adhesives that you can use on jobs such as these. They are mixed according to directions on the package and applied by brush to coat the entire cavity. The adhesive acts much like a wood glue to bond mating surfaces. Often, the adhesive is added to the concrete used for patching as an additional safety factor.

SETTLED SLABS

If a *small* slab has settled in a gridded patio or walk, it is possible that by using one of the two methods shown in the sketches, you may be able to relevel the culprit without having to break up the concrete and replace it with a new pour. Which method you use will depend on the size and weight of the slab and how

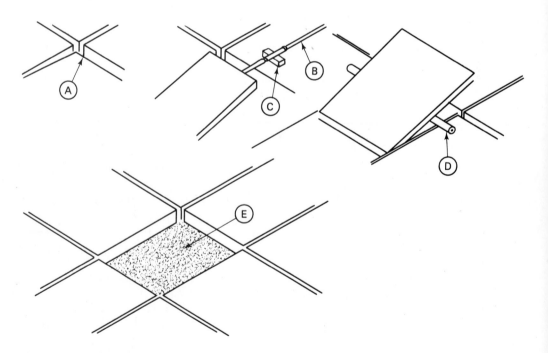

Fig. 25–6 Releveling flagstones or precast slab. Sections may settle, even raise, because of a nearby tree's root growth (a). Pry up the section with a crowbar (b) using a block of wood as a fulcrum (c). Use a length of pipe as a roller to move the section (d). Reorganize the bed before replacing the section (e). If root growth is present it must be removed.

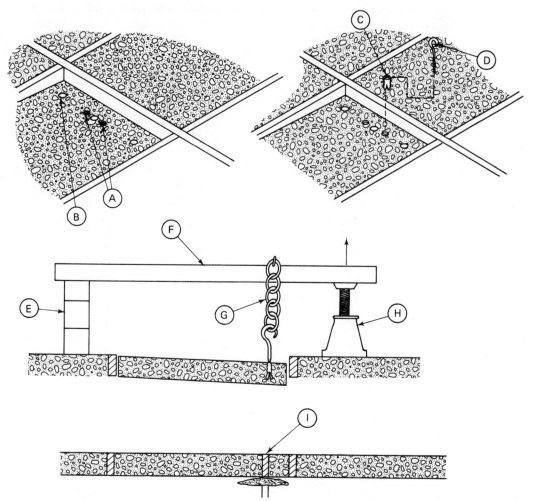

FIG. 25–7   Try this to level a heavy slab, such as a section of a gridded terrace. Drill two holes sized to take a ½ inch expansion sleeve (a). Drill three other holes spaced approximately as shown (b). These should be ¾ to 1 inch in diameter. Drive the expansion sleeves (c) and turn in heavy screweyes (d). Set up two or three blocks as a support (e) for a heavy beam (f). Loop around the beam with a length of chain (g). Use a jackscrew (or an automobile bumper jack) to raise the beam and thus the slab (h). Work through the three large holes with a dowel and hammer to firm the earth beneath and to form a pier hole (i). Work with the dowel to pack a maximum amount of mortar mix through the holes. Allow to set about one week before releasing the beam. Then remove the expansion sleeves and pack the holes with mortar.

it is held in the gridwork pattern. As a guide, you can figure that a large piece of flagstone, especially if it is on a sand bed, will be pretty easy to remove with a roller. If it is a 4-in.-thick piece of concrete 3 or 4 ft² and held tightly in gridwork, the hoist system will be better. If the concrete is tied to the gridwork with nails, you can probably get between the concrete and the wood with a hacksaw blade to cut the bond.

Don't be impatient when you pack mortar through the holes that you drilled through the concrete slab. You want to get as much mortar under the slab as possible. Also, don't use a very loose mix just because it is easier to pass through the holes. Use a regular, heavy-duty mortar mix.

STEP REPAIR

Damage on concrete steps usually occurs at the leading edge of the tread. Repair procedures are the same as those that were detailed for cracks in concrete. Be sure to clean out the damaged section so that the new pour will be keyed to the old. When possible, include reinforcement material, as shown in the sketch. Use the special concrete-adhesive material that we talked about—both as a bonding material before the pour and in the pour itself. Allow plenty of cure time before permitting traffic over the repair area.

When tread bricks loosen, your job is to remove all the mortar that held the brick in place. You can do this by working with a chisel, even an old screw-

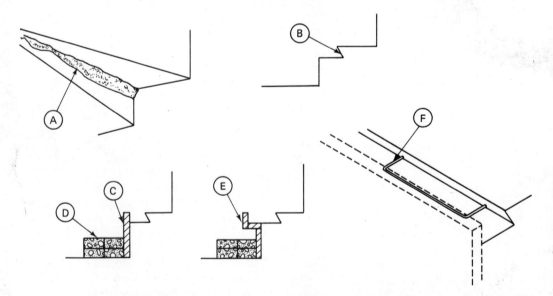

FIG. 25-8 Concrete step repair. Damage usually occurs at the leading edge of the tread (a). Clean out the damaged portion uniformly and be sure to key the back edge (b). When the riser is flush with the tread, set up a form board as shown in (c). Brace the form board with bricks or block (d). When the tread projects, do the form board this way (e). For maximum repair strength, include reinforcement rod—¼ inch—as shown in (f). Rod ends are mortared in holes drilled in the tread. When the patch is small you can work with coat-hanger wire but be sure to sand off any paint.

driver. Take enough mortar out of adjacent joints, and key the bed joint, so that the new mortar you use to replace the brick will tie in to the existing joints. Finish the new joint to match the others. Direct traffic around the repair area until the concrete has set.

TUCK–POINTING

Repair of mortar joints is referred to as *tuck-pointing* or *repointing*. Cracks may appear between mortar and masonry units, which will permit water to pass through a wall. It is also possible for weathering action to cause a joint to deteriorate. In either case, the old joint must be cleaned out and repacked with fresh mortar.

You can work with a small chisel or an old screwdriver—even a spike—to dig out all crumbly and loose mortar. The depth you go depends on the extent of the damage, but the new groove should never be less than ¾ to 1 in. deep.

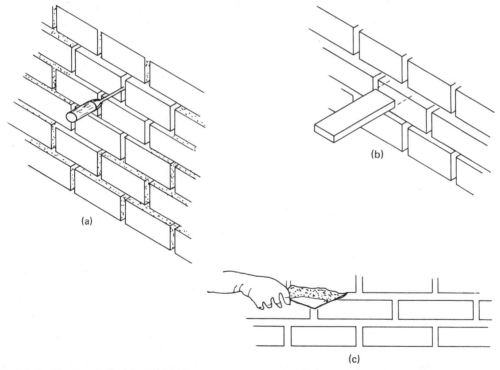

FIG. 25–9   Tuck-pointing is done to repair mortar joint failures. Work with an old screwdriver or small chisel to remove all loose and crumbly mortar (a). The new depth should be ¾ to 1 inch. Moisten the repair area. Use thin wood to repack joints with fresh mortar (b). Finish packing with small trowel (c). Do not tool the joint until new mortar is thumb-print hard.

Work carefully so that you don't crack the brick. Undercut the groove some at each end so that the new mortar will be keyed to the structure.

Use a small brush to clean out the new groove, and check all areas to be sure of minimum depth. Wet the area thoroughly but don't try to do it with a single, soaking application. A few, light hosings, some minutes apart, will do a better job.

The consistency of tuck-pointing mortar should be a little drier than a conventional mix. It is recommended that you mix the dry ingredients thoroughly and then add about half the water that you will need. Let this set about 30 minutes and then add the remaining water.

The best kind of cement to use for this kind of work is *masonry cement* or *plastic cement*. For ordinary service use 1 part of the cement to 2¼ to 3 parts of sand. Heavy-duty types should consist of 1 part of the masonry cement (or the plastic cement), 1 part of portland cement, and 4½ to 6 parts of sand. It is also possible to buy *pointing mortars* in a dry state, ready-mixed. It is always a good idea to have ready-mixes on hand because they provide a convenient supply for small jobs that come up occasionally.

To fill a prepared joint, place the mortar mix on a board and butt the board against the wall just below the joint line. Use a small trowel to slide the mix off the board and into the joint. Work with a square-ended piece of wood to compact and force the mortar into all areas of the cavity. The job will not be right if you leave gaps between the new mortar and the old. Open spaces will trap water and, if it freezes, it will surely push the new mortar outward.

Wait for the new mortar to become thumbprint hard before you tool it to match surrounding areas.

SHRINKAGE CRACKS

Shrinkage cracks can be more unsightly than open or deteriorated mortar joints because they are caused by the separation of units. The crack can follow a joint line, even travel *through* units. If repair work can't be accomplished with routine tuck-pointing methods, the remedial action is to fill the fissure with grout. You can use the same mortar mix recommended for tuck-pointing but with extra water so that it can flow more freely. Use a minimum amount of "extra" water, however; a grout that is too thin will lack strength.

Set up a dam at the lower portion of the crack so that you can pour the grout in layers. The dam may be a braced piece of plywood or even wide adhesive tape. Use whatever means necessary, even a flexible piece of wire as a tamp, to fill the crack. Let the first pour set overnight. Repeat the procedure as many times as necessary to fill the total height of the crack. Be sure that you tool the joints of each pour before organizing for the following one.

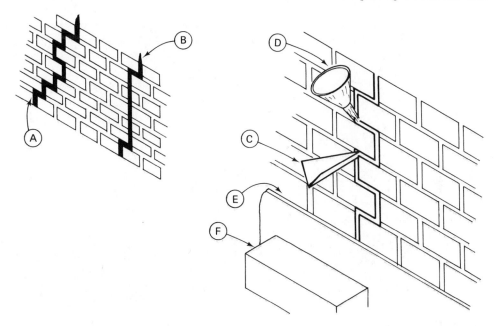

Fig. 25–10 Shrinkage cracks can occur along joint lines (a) or even across units (b). Such cracks should be cleaned and filled solidly with grout. You can work with a V-shaped trough of smooth, heavy cardboard (c), or sometimes you can do it with a funnel (d). Even a length of hose from the funnel may work. Do the job in sections using a piece of plywood to cover the height being filled (e). Brace the plywood cover with blocks (f).

WATERPROOFING A BASEMENT WALL

Interior surfaces of basement walls should be waterproofed even if the exterior surface has a parge coat. This is especially true if you plan to remodel the basement so that it may be used as a family room, recreation room, shoproom, or darkroom. The slightest moisture on a masonry wall can damage a wall covering such as wood paneling.

The extent of the job will depend on the condition of the wall. If it is dry, the chore may consist of no more than an application of waterproof paint. If there are obvious small or large leaks, then these must be sealed before anything else is done.

There are many products on the market designed especially to stop leaks in basement walls. "Waterplug," for example, will stop seepage leaks, and even water flowing under pressure, in minutes. This is the type of product you must use if there is water flowing through mortar joints or through the joint between the wall and the basement floor. It is purchased as a dry powder, mixed with water, and applied like a putty.

FIG. 25–11 Be sure that the entire wall area is thoroughly cleaned. Work with a hoe, trowel, or the back edge of a rake to clean off mortar burrs and anything else that might be sticking to the masonry.

Procedures may differ slightly, but the point is to clean out the joint or the crack that is allowing water to pass through and then to plug the opening with the seal. Use it, too, to plug openings that might be left by tie wires, oversized holes designed for pipes and conduits, and so on.

"Thoroseal" is a cement-base product that is applied like paint. It is a heavy-

FIG. 25–12 All holes, gaps, and cracks must be cleaned out as if you were doing tuck-pointing. If there is water behind the wall, drill weep holes at the base to get rid of the water and to relieve pressure.

FIG. 25–13 Use a product like "waterplug" to completely fill all gaps and cracks and weep holes if any. Mix and apply according to the directions on the container. The material dries and seals quickly.

FIG. 25–14 After the plug material has hardened, you can start with the "paint" job. Most times the masonry must be dampened in order to get a good bond with the cover coat.

FIG. 25–15 "Thoroseal" is made by the same people who make the quick-set "waterplug." It's best to start application at the bottom of the wall, covering the first 3 or 4 courses. Then go to the top and work down.

duty sealant for all masonry and concrete. Use it alone for routine waterproofing; use it *after* applying waterplug when the wall has visible leaks.

All products that fall in this category must be applied very carefully and according to the manufacturer's instructions. All of them will tell you how much of the material to mix with what quantities of water in relation to coverage and the degree of leakage. Sometimes, a heavy base coat is followed by a lighter fin-

FIG. 25–16 The Thoroseal material may be mixed with sand and applied with a trowel or float like a parge coat. Read package instructions for correct proportions. It's also available in different colors.

ish coat. Whatever the instructions are, follow them to the letter if you want optimum results.

A final thought: If the masonry wall is perfectly dry and your interest is merely decorative, it still makes sense to use a waterproofing material. This way you can add a waterproofing safety factor as you achieve the color and the texture that you want.

REMOVING STAINS FROM MASONRY

Dirt stains may be removed by working the area with a wire brush and plenty of water. If the stain is stubborn, try using a detergent or even a scouring powder. Hose the surface generously after you are through scrubbing. Sometimes this will work even with light oil or grease stains.

Hydrated lime or portland cement may be spread like a thick blanket over oil or grease. When this doesn't remove the entire stain, try scrubbing the area with a stiff brush and gasoline or benzine. This should be followed with a good wash, using detergent and water, and then a generous hosing down.

To remove paint stains, use the solvent that the manufacturer recommends as a thinner. Scrub the area with a stiff brush and then wash with detergent and water.

To remove coffee stains, mix 1 part of glycerin with 4 parts of water. Saturate a cloth with the mix and apply it like a poultice to the stain area. Allow the cloth to remain as long as it is needed but keep it wet with the mix.

Wet down blood stains with clear water and then cover with an even layer of sodium peroxide powder. Cover the powder with a water-saturated cloth and allow it to stand for 5 to 10 minutes. Flush with water and then scrub with a brush. You can brush on vinegar (or a 5 percent solution of acetic acid) to neutralize any alkaline traces that remain. The final step is to rinse well with plenty of water. Take all precautions when working with the peroxide powder: Don't breathe it in and don't allow it to come in contact with your skin.

To remove caulking compounds, scrape off as much as possible and then soak what remains with denatured alcohol. This should cause the remaining compound to become brittle enough so that it can be removed with a wire brush. Wash the surface with hot water and soap or clean up with a scouring powder. Usually, caulking compound packages will have instructions for cleanup chores. Follow those instructions before you follow the ones given here.

Chewing gum is tough to remove, but sometimes the same method suggested for removing caulking compounds will work. As a last resort, try soaking with carbon tetrachloride after you have scraped off as much as possible.

To remove efflorescence, mix 1 part of commercial muriatic acid with 2 parts of water. Scrub this into the stain with a stiff brush and flush off with plenty of clear water after three or four minutes. An ammonia solution (about 1 tablespoon to 1 cup of water) will neutralize the acid. Rinse again generously.

CAUTION: *Always pour acid into water slowly. Never pour water into acid.* Also, protect yourself with rubber gloves, eye shields, and the like.

Remove rust by mixing about 1 lb of oxalic acid in 1 gal of water and then applying it generously to the stain. Rinse it away after about an hour or so. Repeat the procedure if the stain is stubborn. Remember the caution about how to combine acid and water.

Do be very careful when using any of the stain-removal systems. Some of the materials are flammable, some are poisonous—most can do some harm if they are handled without due regard for possible consequences. Read and follow all the instructions printed on the package.

Cleaning solutions might even be harmful to nearby plantings. To be safe, build up earth dams so that materials carried away by rinsing will be directed to a safe drainage area or into barren soil.

QUESTIONS

25–1	What is a basic function of *caulking*?
25–2	Name and describe two types of caulking material.
25–3	Name the four ways to buy caulking materials.
25–4	What is an important factor in relation to correct use of a caulking gun?
25–5	Describe a typical procedure for repairing a crack in concrete.
25–6	Describe two methods that can be used to relevel small, settled concrete slabs.
25–7	Make a sketch that shows a typical method of repairing the front edge of a concrete step.
25–8	What is *tuck-pointing*?
25–9	Mortar used for tuck-pointing should be which of the following:
	(A) drier than usual
	(B) a conventional mix
	(C) wetter than usual
	(D) made with cement only
25–10	Is it a good idea to use a special cement for tuck-pointing work?
25–11	What are the special cements called?
25–12	What are *shrinkage cracks*?
25–13	What remedial action can be taken if the crack can't be repaired by tuck-pointing?
25–14	Make a rough sketch that shows how a shrinkage crack can be filled with grout.
25–15	List, in 1-2-3 fashion, the steps to take to waterproof a basement wall.

Answers

1-1   Safer, better highways and walks; sturdier, more attractive homes; the scope of the materials permits such things as backyard swimming pools; the opportunity to have privacy and security without massive structures; etc.

1-2   Flexibility of design; flexibility of texture; flexibility of form; broad application—can be used for small craft projects as well as large construction work; strength combined with beauty; variety of modern materials.

1-3   A design that includes wooden grids; leave open areas for shrubs or trees; special surface textures; perimeter planters for flowers, low shrubs, etc.; perimeter trees; include areas of other materials such as brick, tile, etc.

1-4   It represents the opportunity to visualize individual projects in relation to the total result.

1-5   They indicate the paths used by the residents when moving from one area to another. Also considered are entrances and exits, driveways, areas that should be open to servicepeople, tradespeople, etc.

1-6   They will tell you the type and the size of the project to plan. Walk-widths can be established because of projected use; patio locations can be determined in relation to the home's floor plan; utility areas can be located for convenient use; privacy (or security) considerations will become evident. Also, good planning will help to *direct* traffic.

1–7   It is illegal to ignore the standards that have been established for the job area. You will avoid the errors that led to the establishment of building codes in the first place. Local codes can be a tremendous source of technical advice that applies especially to a specific area. The building inspector can tell you when a variation that you deem necessary to a design, is an acceptable procedure.

1–8   Be sure that the site is ready—check the subbase and the forming; know what must be done when the pouring starts; be sure that you have enough help on hand when it is necessary—be sure that your aides know what procedures will be followed when the pouring starts.

CHAPTER 2

2–1   Incorrect proportioning of the materials required for a mix.

2–2   *Concrete* results when various materials are combined; *cement* is just one of the materials.

2–3   94 lbs.

2–4   When it is combined with water, it makes a paste that surrounds and binds together all the individual pieces of aggregate in the mix.

2–5   The most common *fine* aggregate is natural sand but it can be manufactured by crushing stones or gravel; *coarse* aggregates (usually gravel) can range up to 1½ inches in diameter.

2–6   It should equal about one third (in diameter) the thickness of the slab; when the theory size is not available, one should select the closest, largest size that can be obtained.

2–7   Smaller pieces fill in the empty spaces that would result if only large pieces were used.

2–8   It should be fit to drink.

2–9   The introduction of a material in the mix that results in evenly dispersed microscopic bubbles.

2–10   The water content of concrete can cause enough pressure in a freeze situation to rupture the surface of the material; the tiny bubbles of air that are present because of air entrainment provide relief areas so the pressures will not cause damage.

2–11   It should be bottomless and have inside dimensions that equal 12 inches x 12 inches x 12 inches (one cubic foot).

2–12   *Dry* sand will have no moisture at all; *damp* sand will not compact if you squeeze a handful; *wet* sand will compact when you squeeze a handful but it will not leave the hand wet; *very wet* sand will leave a considerable amount of moisture on your hand.

2–13   Structural failure; chipping; and flaking.

2–14   Buy all the materials separately and in dry state—you do all the proportioning;

buy the aggregates dry but mixed—you add the cement and the water; buy all materials dry but combined (usually sacks)—you add the water; buy all the materials mixed and wet, ready to pour upon delivery.

2–15   (1) Work on a flat surface; (2) spread the correct amount of sand; (3) add the cement and work the two materials together until you achieve a uniform color; (4) add the coarse aggregates and work all the materials together until the blend is uniform; (5) form a hollow in the pile and add the water slowly as you work the material from the perimeter toward the center; (6) continue to mix until all the materials have been thoroughly combined.

CHAPTER 3

3–1   It is a cube that measures 3 feet x 3 feet x 3 feet.

3–2   81 square feet.

3–3   Multiply the thickness of the project by the width and that number by the length. Then divide the result by 12 to find the number of cubic feet.

3–4   Concrete must never be placed on frozen ground; all accumulations of ice or frost must be removed from the site, forms, reinforcements, etc. Correct hardening of the concrete occurs when the temperature of the mix falls between 50° and 70°. If necessary, water, sand, and gravel, should be heated. You must continue to protect the concrete from freezing after it has been placed.

3–5   It assures that little or no moisture will be lost from the concrete during the initial hardening stages.

3–6   Keep the project wet by spraying it frequently with a garden hose; cover the project with special papers or plastic sheeting to keep the existing water from evaporating; keep the project covered with some material like wet burlap; use a commercial spray-on compound to form an anti-evaporation membrane.

3–7   A minimum length of time is 72 hours but keeping it going for as long as a week is a good safety factor.

CHAPTER 4

4–1   A mattock; a short-handled shovel with a square edge; a tamper; and a rake.

4–2   An 8-point, 24 inch utility saw; a 20 ounce framing hammer; an 8 lb. double-faced sledge hammer; a half hatchet; a line; a line level; and flexible tape.

4–3   A short-handled shovel with a square edge and a hoe.

4–4   A short-handled shovel with a square edge and a hoe or rake.

4–5   Screeds or strikeboards—used to level the concrete to the height of the forms; tampers—used to settle the coarse aggregates and to bring a finishable amount of cement paste to the surface; wooden float—starts the actual finishing procedure; bull float with long handle—for areas that can't be reached with a hand float; darby—used like an oversized float; trowel—used after wood-floating to achieve a slicker finish; edgers—to round off the top edges of slabs; groover—to cut control joints; and a cheater—to form decorative grooves.

4–6   Remove all traces of the concrete mix by working with water and a cloth or brush. Wipe dry. Protect them with a light film of oil if they are to be stored.

CHAPTER 5

5–1   It is uniform; compacted to take the weight of the concrete; free of foreign materials; well drained; and dampened to receive the mix.

5–2   Granular fills such as sand, gravel, or crushed stone.

5–3   In thin layers, each layer well-compacted and leveled.

5–4   A sufficient amount of compacting on the surface plus concrete piers down to the original grade.

5–5   Because codes will vary geographically—in some areas it will be safe to place concrete directly on firm soil while other areas require a substantial amount of granular subbase materials.

CHAPTER 6

6–1   In line with the strength requirements of the project. The formwork for a wall requires much more attention than the formwork for a ground slab. The forms must be strong in themselves and braced to support the amount of concrete involved.

6–2   Lumber; plywood; special, knock-down, reusable forms; and textured materials such as resawed or sculptured lumber or plywood for special effects.

6–3   The type of job; the shape of the project; the texture you wish to achieve; and the size of the project.

6–4   A material applied to formwork so that the forms can be removed easily when the concrete has set.

6–5   (The drawing should be done according to the typical layout technique described on pages 41–44.)

6–6   Use a hacksaw to cut at least halfway through the bar and then bend sharply at that point.

6–7   B

6–8   A welded wire fabric that is used most often in concrete slab work.

6–9   Midway in the pour unless specifications suggest otherwise.

6–10   See chart on page 45.

6–11   A maximum of 4 feet.

6–12   A special fastening device that is set in concrete to secure, for example, a wooden sill to the top of a foundation.

6–13   6 feet on centers—at corners and on each side of door openings.

6–14   See illustration on pages 56–57.

6–15   ¼ or ½ inch lumber or plywood; tempered hardboard; sheet metal; stiff fiberglass.

CHAPTER 7

7–1   Place concrete as soon as possible after it has been mixed; dump as close as possible to its final position; do not overwork the concrete when spreading; use a shovel to settle concrete along forms and in corners; tap outside of forms with a hammer.

7–2   (1) Screeding; (2) floating, with a hand tool, a bull float or a darby; and (3) troweling.

7–3   When there is an absence of water sheen and the concrete can take foot pressure without damage.

7–4   *Floating* leaves a slightly rough texture which provides good traction. *Troweling* is done after floating to produce a smooth, hard, dense surface.

CHAPTER 8

8–1   The length of time you allow the concrete to set before you use the broom; whether the bristles of the broom are soft or hard; whether you use the broom wet or dry; whether you use the broom in straight lines, lines that cross, or follow a wave-pattern.

8–2   The *washed finish* is done with water to remove some surface material and so expose some of the aggregate that is part of the mix. *Exposed aggregate* is done by seeding the surface of a pour with special colored material and then washing and brooming the surface so the material is exposed.

8–3   (1) Mix a batch of mortar to the consistency of thick paint; (2) apply it to the concrete surface in a splotchy manner; and (3) allow the mortar to set a short time and then spread and flatten it with a trowel.

8–4   Bend up an 18 inch length of ½ inch copper pipe into a gentle "S" shape.

8–5   Yes. Either by mixing a special pigment with the concrete materials or by working a dry coloring material into the surface of the concrete after it has been poured.

8–6   10 percent of the weight of the concrete.

CHAPTER 9

9–1   An interlocking arrangement of the bricks that provides strength in the structure.

CHAPTER 10

10–1   Face brick; common brick; fire brick; and paving brick.

10–2   SW = brick that may be used where the weather is severe—has high resistance to freeze-thaw and rain-freeze conditions; MW = will take some rain-freeze conditions but not severe ones; NW = for use in mild climates where no freeze or hard frost conditions are possible.

10–3   It is difficult to determine the quality of the product and the pores of the brick may be so filled with old mortar that it will be impossible to get a good, new bond.

10–4   The *nominal* size includes allowances for mortar joints; the *actual* size does not.

10–5   Structural Clay Research—the overall significance of the letters apply to a special masonry that was devised for the trade.

CHAPTER 11

11–1   Trowel; brick hammer; chisel; rule.

CHAPTER 12

12–1   Good design; good materials; dedicated workmanship.

12–2   (1) Draw a 1 inch circle on a surface of the brick that will come in contact with the mortar; (2) place 20 drops of water in the circle; (3) if the water is absorbed in less than 1½ minutes, the brick requires wetting.

12–3   *Stretchers* are placed end-to-end and provide strength longitudinally; *headers* are placed across the wall and provide strength transversely.

12–4   They should comprise not less than 4 percent of the total surface.

12–5   A good rule states that distance between headers either horizontally or vertically should not exceed 2 feet.

12–6   See if it will retain a thumbprint.

12–7   Portland cement; hydrated lime; sand; water.

12–8   As much as is needed to bring the mix to a suitably plastic and workable state.

12–9   One part of Portland cement to ¼ part of hydrated lime to 3-¾ parts of sand.

12–10   It should not be less than 2¼ or more than 3 times the sum of the cement and lime volumes.

12–11   No. In each case, the water should be fit to drink.

CHAPTER 13

13–1   See page 112.

13–2   If possible, this should be done "live" after a demonstration by the instructor. See page 112.

13–3   Spot-placing mortar on the brick and then trying to spread it to form the bed.

13–4   It presents the opportunity to check and to mark off brick placement before you are committed to actual construction. When the design permits, small changes can be made to eliminate or to minimize brick cutting.

13–5   A length of wood that is marked to indicate the position of courses and joints.
13–6   A small error that is multiplied many times can add up to a very big flaw.
13–7   See pages 114–115.

CHAPTER 15

15–1   Sand; crushed stone; gravel; volcanic cinders; expanded slag; specially treated shale or clay.
15–2   8 inches x 8 inches x 16 inches.
15–3   No. The actual dimensions are less than ⅜ inch to allow for the thickness of the mortar joint.
15–4   A unit that has openings through it and which is used mostly for decorative purposes.
15–5   Concave and V-shaped.
15–6   ½ inch square bar stock and ⅝ inch diameter bar stock or heavy tubing.
15–7   1 part of masonry cement plus 1 part of portland cement plus 4–6 parts of sand; and 1 part of portland cement plus ¼ part of hydrated lime plus 2–3 parts of sand.
15–8   No. The block should be kept and used dry.
15–9   Keep them on platforms above the ground. Protect them from moisture, dirt, and dust.

CHAPTER 16

16–1   A *full* mortar bed covers *all* the web areas, while a *partial* mortar bed covers perimeter webs only.
16–2   By placing a *full* bed of mortar on the footing so that you can set down 2 or 3 of the corner blocks.
16–3   On the ends only.
16–4   Build up the corners first, then fill in between.
16–5   By stretching a line between the corner blocks.
16–6   Yes, if local codes dictate that it is necessary.
16–7   Only if it occurs at a corner.
16–8   A control joint should be used at that point. Tie-bars spaced not more than 4 feet apart (vertically) should be included in the joints.
16–9   In this case, pieces of metal lath are placed across the joint that connects the two walls. The lath should be placed in every other course.
16–10   By placing a piece of metal lath over a core that is two blocks down from the top. This makes it possible to fill the core with concrete so the anchor bolt can be placed.
16–11   About 4 feet on centers.
16–12   So that movements caused by various kinds of stresses can occur without doing damage to the structure.

16–13   See pages 155–156.

16–14   A system of applying a portland cement paste to a block wall to waterproof it.

16–15   ½ inch thick.

16–16   By cutting with a chisel or a masonry-cutting saw.

16–17   Wearing safety goggles for eye protection.

CHAPTER 17

17–1   Good traction for safety.

17–2   3 to 4 feet wide for entry walks; 2 feet wide for service paths.

17–3   About ¼ inch per foot of width. Unusual conditions may require ½ inch per foot.

17–4   4 inches thick.

17–5   A groove cut in concrete so that a crack will occur, ideally, under the groove where it will not be visible.

17–6   It eases potential stresses where a slab abuts a house or a driveway or an existing concrete slab.

17–7   It is used when a pour must be done in sections and the design does not include permanent dividers. It is included so that abutting slabs will interlock and maintain a level surface.

17–8   See pages 170–171.

17–9   Over a concrete subbase, and over a bed of sand.

17–10   See pages 172–173.

17–11   See page 174.

17–12   Fill the joints with sand by brooming; fill the joints with wet mortar; fill the joints by brooming in a dry mortar mix and then wetting it down by spraying with a garden hose.

17–13   Flagstones and patio tiles.

17–14   A full 4 inches will do for passenger vehicles but 5 to 6 inches is more appropriate for occasional truck traffic.

17–15   10 feet.

17–16   Then the minimum width should be 14 feet.

17–17   Its width should be 3 feet wider than the cars it must accommodate.

17–18   See pages 178–179.

17–19   Avoiding changes that will cause scraping of the undersides and bumpers of automobiles.

17–20   Include a post for a basketball backstop; posts for a volley ball net; a shuffleboard court; layout for hop-skotch; etc.

CHAPTER 18

18–1   The thickness range is from 1½ inches up to a full 4 inches. It depends on the size, use, and placement of the stone.

18–2   In special forms that are made for the purpose; by digging the form directly in the ground; by making a sidewalk-type form with divider strips that remain in place after the concrete has been poured.

18–3   By using the corrugated sheet metal (or plastic) that is sold for lawn edging as a mold.

18–4   See page 188.

18–5   When a casual path is needed across an existing lawn.

18–6   By using a very plastic mortar mix as a topping.

18–7   Cardboard boxes; hoops; plastic containers; parts of wooden crates; etc.

18–8   Ideas should be practical but imaginative.

CHAPTER 19

19–1   Outdoor cooking; sun bathing; outdoor dining; reading; games; etc.

19–2   A northern exposure will get minimal direct sun; southern exposure will get the most sun; western exposure will get the most sun in the afternoon; eastern exposure will have cool afternoons.

19–3   It makes a large expanse of concrete look more attractive; it divides the pouring job into a series of small slabs; it's easy to leave open spaces for plantings; the gridwork itself can be designed attractively; different materials may be combined.

19–4   ⅛ per foot of run.

19–5   The nails act as a tie between the wood and the concrete.

19–6   See page 203.

19–7   They should be cut off at about the center of the runners because they are a permanent part of the assembly and must be covered by the concrete.

19–8   *Nominal* dimension bricks are designed for use with a mortar joint—*actual* dimension units are not.

19–9   See pages 208–210.

19–10   Perimeter bricks can move out of place unless they are set solidly.

CHAPTER 20

20–1   See page 223.

20–2   See page 224.

20–3   Line the inside surfaces of the forms with materials like plastic sheet, roofing felt, and kraft paper.

20–4   When the wall can't be done in a single pour. See pages 225–226.

20–5   The radius of the curves should not be more than 2 times the above-grade height of the wall; the depth of the curves should not be less than ½ the height.

20–6   The above-grade height of the wall should not be more than ¾ the wall-thickness squared.

20–7   Considers only wind and impact loads; assumes no special bond between the wall and its foundation.

20–8   See page 230.

20–9   A pattern that leaves open spaces between the stretcher bricks.

20–10   One part of portland cement plus ¼ to ½ parts of hydrated lime plus 2¼ to 3 times the total of the cement and the lime.

20–11   1 part of portland cement plus 0–1/10 part of hydrated lime plus 2¼ to 3 times the total of the cement and the lime.

20–12   When the cavity in the wall is 2 inches wide or less.

20–13   When the cavity in the wall is more than 2 inches wide.

20–14   ⅜ inch maximum size coarse aggregate—1 to 2 times the total of the cement and the lime.

20–15   Mainly, to keep excess mortar from falling into the cavity and mixing with the grout.

20–16   See page 232.

20–17   Cast-in-place concrete or mortar; wood; brick; flagstone; tile; readymade coping.

CHAPTER 21

21–1   See page 246.

21–2   If the slope is very shallow or if the traffic will include wheeled vehicles such as lawn mowers, wheelbarrows, bicycles, etc.

21–3   See page 254.

21–4   By using rubble as fill material inside the forms.

21–5   A concrete subbase in the form of steps is poured first and the brick is used as a veneer; the brick steps are constructed on a solid concrete slab; concrete is poured as a perimeter footing only.

21–6   SW brick and type M mortar.

21–7   See pages 263–264.

CHAPTER 22

22–1   Anchor bolts; brackets for posts; steel plates for posts and similar items; concrete inserts.

22–2   Toggle bolts; "molly" bolts; picture frame hooks; expansion sleeves; anchors.

22–3   A fastener for a *hollow wall* becomes secure when part of the fastener expands and grips against the blind side of the wall. A fastener for a *solid wall* is designed to expand and grip the sides of a hole that is drilled for it.

22–4   A special tool that is designed for driving with a heavy hammer to form holes in masonry materials.

22–5   They should be carbide-tipped.

22–6   ½ inch because it will have enough power, and suitable speeds.

22–7   Wear gloves; wear safety goggles; don't allow anyone to stand nearby; use a heavy hammer or a small sledge.

CHAPTER 23

23–1   Because it will be difficult to keep from forming an oversize hole. This calls for excessive amounts of concrete and disturbs too much soil.
23–2   A clam-shell digger and an auger. It is easy with either one to form a hole that is about 6 inches in diameter—an ideal size for 4 inch x 4 inch fence posts.
23–3   See page 288.
23–4   Use white portland cement and white aggregates in the mix.
23–5   See page 292.
23–6   See page 297.
23–7   A five-gallon metal drum and a discarded automobile tire.

CHAPTER 24

24–1   Vermiculite and haydite.
24–2   1 part of cement plus 2 parts of sand plus 3 parts of vermiculite.
24–3   By adding about ¼ part of hydrated lime or fire clay.
24–4   Crushed walnut shells; styrofoam fill; marbles; glass beads; wood chips or sawdust.
24–5   See page 312.
24–6   Answers should be judged on degree of creativity and methods used for casting.
24–8   See page 320.
24–9   It takes water from a pool and pushes it up to a higher level from where it drops down into the pool. No additional water is required to keep the pool full.
24–10   See page 324.
24–11   By installing a regulator or by placing a restrictor on the hose.

CHAPTER 25

25–1   To oppose the penetration of air and water.
25–2   See chart on page 335.
25–3   In pre-loaded cartridges; in bulk form; in rope form; in small squeeze-out tubes.
25–4   To synchronize the amount of material that is squeezed out of the tube with the speed at which you lay down the bead.
25–5   See page 337.
25–6   See pages 338–339.
25–7   See page 340.

25–8   The work required to repair a mortar joint that has failed.

25–9   A

25–10   Yes.

25–11   Masonry cement and plastic cement.

25–12   Cracks that are caused by the separation of units.

25–13   The fissure should be filled with a grout.

25–14   See page 343.

25–15   (1) Clean the wall thoroughly; (2) clean out any existing holes, gaps, or cracks;
         (3) if necessary, drill weep holes at the base of the wall to relieve water pres-
         sures; (4) use a product that is made for the purpose to fill gaps, weep holes,
         cracks, etc.; (5) allow the plug material to harden; (6) coat the wall with a spe-
         cial waterproof paint that is made for the purpose.

Index